Automated Weighing Technology

Automated Weighing Technology: Process Solutions

Ralph Closs
Henry Vandelinde, Ph.D
and Matt Morrissey

MP MOMENTUM PRESS

MOMENTUM PRESS, LLC, NEW YORK

Automated Weighing Technology: Process Solutions
Copyright © Siemens Canada Limited, 2013

Published by:

Momentum Press®, LLC
222 East 46th Street
New York, NY 10017
www.momentumpress.net

ISBN-13: 978-1-60650-633-2 (hardcover, casebound)
ISBN-10: 1-60650-633-1 (hardcover, casebound)
ISBN-13: 978-1-60650-634-9 (e-book)
ISBN-10: 1-60650-634-X (e-book)
DOI: 10.5643/9781606506349

Cover design by Jonathan Pennell

10 9 8 7 6 5 4 3 2 1

Printed in the United States of America

Dedication

As always, a book like this is the concerted effort of many people both present and past whose experience and input contribute to these pages. From application engineers to designers to promoters to sales people to customers to the folks that work diligently to assemble the components with care, the quality of Siemens weighing technology is a testament to their dedication and pride.

The authors begin by thanking our families who stood by while we poured over the manuscript on weekends and nights. Your patience while we obsessed over the minutia that mattered only to us is much appreciated and we love you letting us do so. Crystal, Will, Sarah and Harry thanks for pretending weighing is interesting. We also thank Cam Clements for his contributions and Graham McGregor whose passion for weighing was instrumental in getting this project off the ground. To Martin Schoettinger who reviewed the manuscript – you have our appreciation – and to all of you who provided advice and technical input we extend a heartfelt thank you.

As with the other Siemens Milltronics books, Pete Froggatt lent his creative talents to the graphics while Jamie Chepeka kept everything moving. Dana Van Allen's critical review put the sparkle in and Carla Guest's eagle copywriting eye kept us from looking too foolish.

To all of you, the best of this book is yours, and any mistakes and omissions are solely the responsibility of the authors.

On a final note, we all want to extend our appreciation to the one voice that stands out in these pages – that of Ralph Closs. His is a lifetime spent in this field, witnessing four decades of technological change and market growth while steadily advancing Siemens weighing. Ralph's experience is second to none and this book is in many ways a snapshot of a career dedicated to a field that is better for having him in it. As a colleague, Ralph brought invaluable experience and as a writing partner he added wisdom and charm. Most of all, to all of us here at Siemens Milltronics, he was and remains our friend.

This book belongs to you Ralph.

Foreword

Weighing technology by its very nature suggests something solid and substantial, which is exactly what you get from Siemens – and a whole lot more. From the rugged requirements of a mine to the delicate measurements of pharmaceutical suppliers, Siemens weighing technology provides accurate weight measurement whether the material is on the go or standing still.

The unmatched performance of the Siemens Milltronics Single Idler (MSI) leads the way, supported by the whole Siemens line of belt scales, weighfeeders, and flowmeters. These contact instruments provide the data which is then interpreted by Siemens integrators like the SITRANS BW500 and the SIWAREX FTC module.

Siemens weighing solutions are supported by peripheral devices and equipment like speed sensors, test chains, and the Milltronics Weight Lifter (MWL). Furthermore, solutions are often custom designed by Siemens technical staff and application engineers to ensure optimal performance.

Siemens provides complete weighing solutions for many industries from food and beverage to pharmaceuticals and environmental systems. However, the mining, aggregate, and cement industry is a primary focus where Siemens rugged reliability and accuracy play a key role. When experts ask about the best in the weighing field, the answer is clearly Siemens.[1]

Keywords

belt scale, weighing, belt weigher, load cell, speed sensor, MSI, BW500, WS300, milltronics, flow rate

---

[1] For more on Siemens weighing technology, including products, documentation, and training, go to www.siemens.com/weighing. For videos on Siemens weighing products, go to the Siemens YouTube channel: www.youtube.com/user/thinksiemens.

Contents

Chapter 1

Introduction to Weighing

When heavy is the weight of the world.[1]

We are a weight obsessed world. From our own personal poundage to the latest celebrity diet, we scrutinize and make value judgments based on weight. Without delving too deeply into the cultural and psychological implications of this, suffice it to say that our concerns over our personal weight reflect a necessary concern with weight that is a cornerstone of our economy. We buy things by weight, and we value things by weight (on both ends of the spectrum), and the requirements of fair barter and trade necessitate an accurate under-standing of weight—accuracy that is at the heart of Siemens weighing technology.

This chapter draws a brief outline of the development of weighing mechanisms through the ages, from simple balance scales to the sophisticated systems Siemens brings to the market. Siemens has been in the weighing business since the 1970s and has set the standard for belt scale weighing, reducing complicated and clunky multi-idler systems to sleek and reliable single-idler units feeding data through a sophisticated digital control system.

Topics

- Types of weighing
- Terminology
- Weighing history

Types of weighing

There are two types of weighing processes: static and dynamic.

Static weighing

Static weighing refers to when material is placed on a weighing mechanism (platform or vessel) and then typically sits at rest while being weighed. This static form is the most common weighing

---

[1] Van Halen, *Dirty Water Dog*. 1998.

reference and permeates our everyday lives, from our homes to the doctor's office, grocery store, and just about any factory, manufacturing, research, processing facility, or office on the planet.

Types and styles of static scales vary from the very generic to those suited for specific and precise applications.

Dynamic weighing

Industrial applications often require the ability to measure and weigh material in motion for the loading of vessels or feeding into production. Production processes rely on the movement of material, and stopping to weigh means stopping production, a costly exercise. Therefore, material must be weighed in motion, without stopping its flow or temporary storage.

Dynamic weighing devices are scales that quantify bulk solid materials as they are conveyed from one point to another. Among these dynamic devices are belt conveyors, belt feeders, gravity fed chutes, and pneumatic conveyors.[2]

Weighing history[3]

As civilization developed with barter and trade, it brought along the need to measure, count, and weigh. Weighing devices have been developed in every size and shape imaginable to suit countless applications over the years to weigh everything from seeds to livestock, from zeptograms (10^{-21}) to gigagrams (10^9), and from a single protein molecule to an ocean liner.

It is thought that the first balance type of weighing device was used during the Mesopotamian age of civilization more than 6000 years ago near the advent of the Bronze age. Based on the balancing principle, these early devices were thought to have been made from wooden, or even stone, beams suspended in the center with supporting wooden containers at the ends. One end held the goods to be weighed, the other the reference weights.

Over the centuries these balance beams were refined and improved. They were constructed of better materials and longer lasting pivot points were added as lever ratios evolved into more complicated mechanics. The development of the reference masses also went through many evolutions and attempts to standardize.

---

[2] See *Chapter Three* for more on dynamic devices.
[3] This brief outline is intended to set the context for Siemens industrial weighing only.

Balance scales dominated weighing until the nineteenth century when the first indicating scales were introduced. Although conceptualized by Leonardo Da Vinci in the fifteenth century, indicator scales that identified the value of the material as well as its weight did not transform the commercial weighing business until the 1800s.

These scales used levers and springs to provide the required resistance and corresponding value indicators. This type of mechanical weighing dominated the industry because it could offer a precise measurement not possible with older devices that relied on shifting counterweights and proper placement.

Static industrial weighing

Mechanical scales dominated as bins, hoppers, silos, trucks, and large goods were weighed mechanically, often on platform scales, with the values shown on indicator dials similar to a bathroom scale readout. In the late 1940s, the development of the load cell, which converted mechanical forces into an electric signal, changed the industry. Hoppers could be placed on load cells for batching and mixing applications, and load cells make check weighing and totalizing easier.

When strain gauge load cells and LVDT load cells became available, most dial indicators were eventually replaced. Amplifiers/indicators were added to provide load cell excitation and to provide an interface with early control systems.

Amplifiers/indicators are still used in some applications for local readout, but the norm now is to connect to control systems using PCs, DCS, or PLC devices, either via analog circuits or by using industrial bus communications.[4]

---

[4] For more on industrial communication, see *Catching the Process Fieldbus*. James Powell and Henry Vandelinde. Siemens, 2009.

Conveyor belt scales

Herbert Merrick is credited with developing the first dynamic weighing machine in 1908.[5] For the first time in industrial weighing history, it was possible to weigh the material being conveyed while it was in motion. These early machines monitored production rates and estimated inventory on their own without requiring surveyors, increasing production and reducing costs.

Mechanical scales would evolve until the 1960s when strain gauge load cells and electronic circuits brought in the electronic age of weighing. Load cell technology improved continuously, and the introduction of the microprocessor in the 1970s changed the face of the original mechanical belt scale forever.

As the newer, faster, better belt scale designs evolved, industry benefited from greater accuracies and repeatability. Accuracy is crucial, but repeatability is also important to trade as it provides confidence in the measurement. The scales assumed significant responsibility by improving accountability, process efficiencies, and production quality; and by providing cost-effective measurements legal for trade.

With the growth of industry comes the requirement for higher accuracy in production and packaging, and belt scales provide that function. Typical belt scale applications include the following:

- monitoring production and the flow of raw materials
- feeding processes, including batch feeding and flow rate control
- receiving, stockpiling, shipping, and packaging
- mixing, quality control, pollution control
- recycling of materials, waste management
- inventory control and monitoring
- load-out optimization to truck, rail, and ship holds; and custody transfer measurements

Almost every dry bulk solid material that can be conveyed on a belt can be effectively weighed by a belt scale. Belt scales can be installed on conveyors that are indoors, outdoors, underground, on-board vehicles or vessels, and in practically any climate or part

---

[5] Merrick went on to establish Merrick Scale Manufacturing Company, now owned by Tannehill International Industries, and Merrick is still manufacturing scales. See http://www.merrick-inc.com/index.html?/AboutUs/History/index.html for more information on the history of Merrick scales.

of the world. Available in all shapes, sizes, and construction materials, they can measure as little as a few kilograms per hour of operation and as much as thousands of tons per hour.

Terminology

The following are central to a clear understanding of dynamic weighing applications and the systems employed.[6]

Balance

A weighing device usually comprising a rigid beam horizontally suspended by a low-friction support at its center. Identical weighing pans are hung at either end; one pan holds an unknown weight to be determined while the other pan has known weights added to it until the beam is level and motionless. The unknown weight is then calculated as the sum of the weights added to the other pan.

Gravimetric feeding

A feeding system that uses a transducer[7] such as a strain gauge load cell or an LVDT to measure the results of gravitational force exerted by material on a conveyor belt and a signal proportional to the speed of the conveyor belt to derive material flow rate. These devices are typically reliable, repeatable, and accurate and are not affected by the material factors that can affect the volumetric systems because they are feeding based on a measured mass and not only on volume.

Load cell

A transducer that converts force into an electric signal. A strain gauge load cell measures the strain on the load cell as an electrical signal.

LVDT

Linear variable differential transformer, an electrical transformer that measures linear displacement.

Metrology

The study of weights and measures, including units of measurement.

Scale / Weigher

An instrument, device, or machine for weighing.

---

[6] For a complete list see *Appendix A: Glossary.*
[7] A device or sensor that converts one form of energy into another.

Volumetric feeding

A feeding system using a device to deliver a specific, predictable and repeatable volume of material with each revolution or cycle. As long as the material being delivered is uniform in size, mass, moisture content, and bulk density and is not prone to bridging between surfaces or sticking to surfaces, these devices can be varied in their speed to deliver a specific rate of material feed. As the material factors become less uniform or stable, the reliability and repeatability of these feeders decreases.

Weigh (v.)

To determine or ascertain the force that gravity exerts on a balance, scale, or other mechanical device.

Weight

The amount something weighs using a known reference. It is a measurement of the gravitational force on a body, equal to the mass of the body multiplied by the acceleration of gravity.

Weights and measures

The laws, acts, rules, governing bodies, agencies, or standards dealing with the science of metrology and the measurement of physical quantities.

Industrial weighing and Siemens

Siemens Milltronics

Milltronics was founded in 1954 in Peterborough, Ontario, Canada before becoming part of Siemens in 2000. Starting with grinding mill control systems using detection of sound frequency of ball mills, Milltronics grew into a global supplier of these systems to the mining and cement industries. During the late 1960s and early 1970s other sensor based products, such as motion detection, for the heavy industries were added to the product line.

In the 1970s, Milltronics expanded its product offering with the following:

- ultrasonic sensors when it acquired Raytheon's industrial ultrasonic business[8]

---

[8] For more on the development of the Siemens Milltronics ultrasonic product line, see *Ultrasonic Level Measurement*, 3rd Edition. Stephen Milligan and Henry Vandelinde. Siemens, 2012.

- marketing Sankyo Impactline[9] solids flowmeters in the Americas after reaching an agreement with Sankyo Dengyo
- manufacturing and marketing of Merrick belt scales and weighfeeders in Canada after reaching an agreement with Merrick Scale of Passaic, New Jersey. This agreement was in effect from 1975 to 1980.

These products not only served the existing targeted cement and mining markets, but helped grow business in other primary and secondary industrial markets.

The Merrick years

The belt scale first developed by Merrick Scales in 1908 used a mechanical design that integrated the load and speed measurements for totalization and flow rate determination.[10] This mechanical belt scale and integrator design was very popular and quite accurate but required a great deal of attention. Its fast wearing parts needed considerable servicing time, replacement components, and re-calibrations to maintain the required performance.

These belt scales also required a number of suspended idlers (2, 4, 6, or even 8) for the following reasons:

- guarantee sufficient load forces
- compensate for idler misalignment from mechanical deflection of the scale frame and integrator movement as the conveyor is loaded
- provide enough retention time on the weighbridge to allow the full transfer of forces

I met an instrument technician who worked at a cement plant that had many old mechanical weighfeeders. We chatted and he told me he was also a watchmaker. Although he was very pleased to have been hired over many other applicants, he soon realized that his watchmaking experience was certainly the main reason he was hired.

Apparently, calibrating, repairing, and maintaining many of these aging weighfeeders required the same attention to detail and finesse.

---

[9] Copyright Sankyo Dengyo corporation.
[10] See *Chapter Seven* for more detail on this integration.

Eventually, electronic devices replaced the mechanical integrator. Strain gauge load cell designs provided greater levels of accuracy and reliability, giving belt scale manufacturers the opportunity to develop simpler belt scale designs. The load cells provided analog mV output which could be combined with the separately measured belt speed. An electronic integrator processed these two values to determine the totalization and rate.

Merrick developed an analog-based integrator to calculate weight values and then went on to an integrator with limited digital processing. The belt scales used were generally multi-idler devices. The exception is the single-idler version reserved primarily for weighfeeders and weighbelt conveyors.

After the Merrick-Milltronics agreement ended, Milltronics designed the MIC, its own belt scale with multi-idler versions as well as a single-idler unit for weighfeeder and weighbelt usage. The first integrator developed by Milltronics, the SA (Scale Amplifier), was an analog based device with limited digital processing.

Milltronics belt scales were well received in the traditional markets but there was a growing demand for lower cost, yet accurate, belt scales. The market was also looking for an easy-to-use belt scale integrator that would simplify setup and calibration, negating the need for running a number of belt circuits over the belt scale and making best guess adjustments.

In the late 1970s, Milltronics offered the first full integrator solution with the CompuScale®,[11] the first micro-processor based integrator. The market snapped it up, and customers demanded it be supplied with new belt scale systems. It worked so well, many customers developed specific maintenance schedules to replace existing integrators in the field with the CompuScale, regardless of the original manufacturer of the installed system.

As the CompuScale revolutionized the integrator portion of the belt scale business, Milltronics engineers based in Arlington, Texas were busy revising the traditional belt scale design. They took the unique

---

[11] CompuScale is registered trademark of Siemens Milltronics Process Instruments.

parallelogram single-point load cell Milltronics already used in the design of its M series solids flowmeter and used it in a belt scale, creating the MSI. The MSI single-idler belt scale with its active load cell summation, revolutionized belt conveyor weighing by offering a truly accurate, compact single-idler belt scale.

SIWAREX

Although Siemens Process Instruments, based in Karlsruhe, Germany, experimented with the manufacture of weighfeeders in the 1970s in order to supply some large turn-key projects, its weighing focus has largely been in the integration end of the belt scale, weighfeeder, and weighbelt business of dynamic weighing.

Using Siemens PLC modules as interface, the load cell signal and speed sensor signals were processed by a Siemens PLC, acting as integrator to establish the rate of material and the totalization of the same. Siemens created its weighing group – SIWAREX – in the 1970s to develop PLC based weighing modules specifically for accurate excitation of the load cells and high accuracy processing of load cell signals. Siemens also designed load cells specifically for manufacturers of level by weight, process batching, and continuous process control.

The SIWAREX weighing group has also developed a PLC based weighing module as a direct interface with belt scales, weighfeeders, and weighbelts. This FTC module (Flexible Technology for Continuous Weighing) launched in 2004 performs the load and speed integration directly within the module instead of the main PLC control software. The FTC is considered a critical milestone in the conveyor belt weighing business and is now also used with solids flowmeters, batch weighing, and loss-in-weight continuous rate control.[12]

Other SIWAREX components like the U+CS are used for basic weighing applications like platform scales and gravimetric level management. The FTA is used in batch processing, check weighing, and for totalizing scales.

---

[12] At time of publication, the SIWAREX WP241 belt scale module in the S7-1200 PLC family is in development. With new stand-alone functionality, it can be used with any PLC system or directly with an HMI. The module has four digital inputs and outputs, as well as a 4-20 mA output. More importantly it has Modbus TCP I/P and RS485 communications as standard features.

Summary

Weighing has an integral role in the Siemens process control business. Its long established presence in the marketplace and the reputation it has built firmly establishes Siemens as a global weighing supplier. By combining the Milltronics belt scale designs and stand-alone belt scale integrators with Siemens SIWAREX PLC based weighing electronics, Siemens provides answers to the toughest questions in both dynamic and static weighing.

The engine that drives Siemens dynamic weighing is the MSI belt scale. Its unique design, rugged capabilities, ease of use, and dependable performance place it at the center of the weighing line. Integrators, speed sensors, load cells, weight lifters, and other components make up the complete weighing package.

This book presents the Siemens weighing line, with emphasis on the belt scale and its configurations, applications, and many advantages. So sit back and enjoy the show as we show you how to measure our *weigh*.

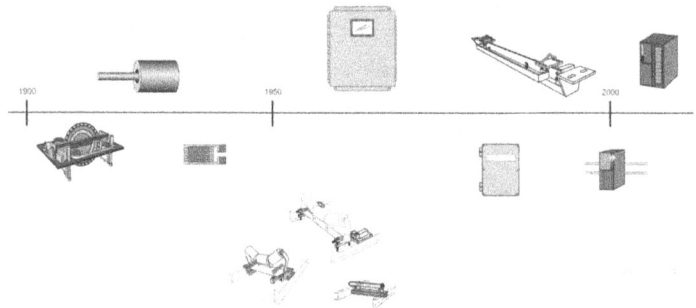

Chapter 2
Bulk Material Handling

Not everyone can carry the weight of the world.[1]

The weighing process generally comprises two types of devices – one to move the material and one to measure it. In the interest of linearity, the transport mechanisms are discussed in this chapter and where they are best applied, followed in Chapter Three by devices that weigh the material on the move.

Moving dry bulk solids from one location to another can be accomplished numerous ways. While the concept seems simple enough – take this stuff and move it over there – actually getting it there is a bit more complicated. The material can change elevation or be extracted from, or into, a container or stockpile. It can be readied for transport or stored in a silo. It may be part of a process or feed a production requirement where the handling has to be modulated; or it may need to be accurately controlled or precisely measured.

Material type, as well as where it comes from and where it is going, often determines the handling process and device. This chapter presents the most common handling devices available on the market. A quick application reference chart follows each section outlining typical applications where the device is used.

For reference please note the following indicators:

- **Yes** indicates that the feeder has been used in that industry
- **No** means that it has not been used or recommended in that industry
- **Possible** indicates it is technically feasible, but we are not aware of actual applications in the specific industry

---

[1] REM, "Talk About The Passion." *Murmur.* 1983.

Belt feeders

Belt feeders shear material out of a bin and feed into a process on a volumetric basis, with the material depth on the belt set by a vertical gate. Designed for high shear loads and resulting high belt tension, belt feeders generally comprise the following:

- a head pulley driving the belt
- rollers or idlers supporting the belt
- a rotating tail pulley that allows the return strand of the belt to go back and become the upper conveying strand

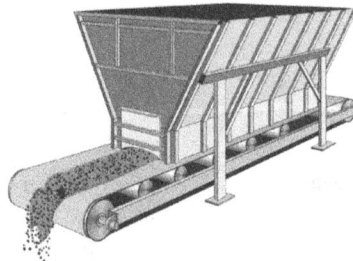

This volumetric device can be modified for gravimetric weighing control. This mode requires there to be sufficient distance between the material in-feed section and the discharge point for a belt scale to be installed. The belt scale supports one or two conveying rollers which are in turn solely supported by the load sensing component of the belt scale.

These weighing idlers float in reference to the rigid rollers, supported by the "weighbridge" of the belt scale and reacting to the weight of the belt. That reaction is used in the feed rate calculation along with the belt speed which is measured by a speed sensor typically mounted on the tail pulley shaft.

It's a misconception to think that installing a belt scale on a conveyor gives you a conveyor system with a weighing device. In fact, the conveyor becomes part of the weighing system, and anything that is changed, repaired, adjusted, or maintained on a conveyor system affects the performance of the weighing system. When I inspect a belt scale system, I walk the length of the whole conveyor.

I am looking, listening, feeling, and smelling (I try not to do too much tasting). I look for obvious damage or failed components, missing components, tramp material, recent repairs, shiny or recently painted components or sections. I listen for squealing, scraping, grinding, and whopping or rubbing sounds. I feel for vibration, extreme temperatures, or strong air movements; and I sniff for burning motors, burning or slipping drive belts, burning rubber belting, or even failing bearings. This physical inspection indicates a lot about the application and the quality of the attention paid to the weighing system.

Belt feeder application	
Mining	Yes
Aggregate	Yes
Cement	Yes
Food	Yes
Chemical, fertilizer (dry)	Yes
Grain	Possible. Not common
Steel and coal	Yes
Water/wastewater	Yes

Belt conveyors

Belt conveyors move material from one location to another without shearing from a bin. They are loaded from one or more feed points by devices such as apron feeders, belt feeders, vibratory feeders, screw conveyors, table feeders, rotary feeders, or even by hand.

Belt conveyors are similar in design to belt feeders, comprising the following:

- a head pulley which drives the belt
- rollers or idlers that support the belt
- a tail pulley that rotates to allow the return strand of the belt to come back to serve as the upper conveying strand

Because belt conveyors are generally longer than belt feeders, they often require return idlers to support the return strand of the conveyor belt. Installing a belt scale on a belt conveyor is generally easier than on a belt feeder as there is more available space.

Sometimes a short belt conveyor is used primarily to weigh the material in a process, with transferring material a more secondary requirement. This arrangement is often referred to as a weighbelt.

Belt conveyor application	
Mining	Yes
Aggregate	Yes
Cement	Yes
Food	Yes
Chemical, fertilizer (dry)	Yes
Grain	Yes
Steel and coal	Yes
Water/wastewater	Yes

Screw feeders and conveyors

Screw feeders and conveyors are often used as pre-feeders to dynamic weighing devices. A variable speed controller added to the drive system will modulate the volumetric flow rate of the material to the weighing device, allowing for process control with the dynamic weighing element acting as the measured process variable.

Screw feeders: these feeders shear material from a bin and can volumetrically modulate the flow of material. The smaller variety of screw feeders are very precisely designed and built for such purpose. To change a volumetric screw feeder to a gravimetric design, the smaller units and their pre-feeding bin can be supported by precision load cells and special controllers with software functions to create a loss-in-weight system.

Screw feeder application	
Mining	Yes
Aggregate	Possible. Not common
Cement	Yes
Food	Yes
Chemical, fertilizer (dry)	Yes
Grain	Yes
Steel and coal	Yes
Water/wastewater	Yes

Screw conveyors: conveyers move material from one location to another, usually without shearing action. A weigh screw is often used when a totally enclosed, pressure-rated conveyor with dynamic weighing is required. The weigh screw operates by supporting the screw conveyor with load cells and by using flexible in-feed and discharge transitions which limit interference of the weighing process by the system chute-work.

The screw shaft speed is also monitored, extrapolating a linear conveying speed. The loading and speed signals are integrated for the calculation of instantaneous rate and accumulated total. A belt scale integrator is often used for this function.

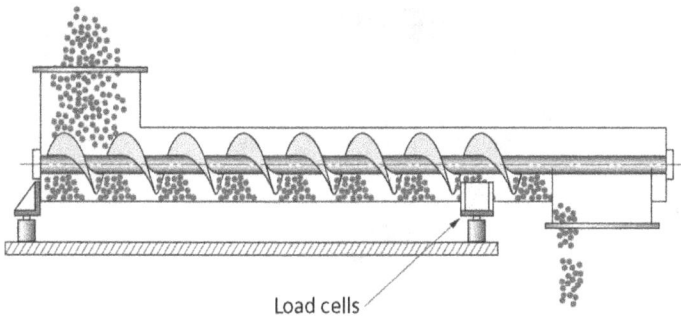

Load cells

Screw conveyor application	
Mining	Yes
Aggregate	Possible. Not common
Cement	Yes
Food	Yes
Chemical, fertilizer (dry)	Yes
Grain	Yes
Steel and coal	Yes
Water/wastewater	Yes

Apron feeders

Apron feeders shear large course materials from dump hoppers which are often filled by large loaders. These feeders are built with interlocking steel pans supported and interconnected by a heavy chain on each side. The chains are supported by custom-designed bogey wheels.

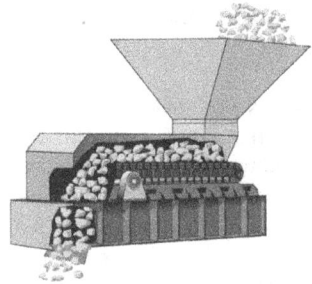

The pans have cleats similar to bulldozer tracks to add to the material traction and to aid in the extraction of large pieces of material from the bin or hopper. This feeder often supplies crushers, breakers, or conventional belt conveyors.

Apron feeder application	
Mining	Yes
Aggregate	Yes
Cement	Yes
Food	No
Chemical, fertilizer (dry)	No
Grain	No
Steel and coal	Yes
Water/wastewater	No

Pan feeder

Pan conveyors are similarly constructed to apron feeders but are used to convey material from one location to another without shearing from a bin. Pan conveyers are also used to move hot materials that cannot be transported on rubber conveyor belts.

They are built with interlocking steel pans supported by bogey wheels running on an iron construction track similar to rail road track. The pans are cleated (similar to bulldozer tracks) to assist moving the material, often on sharp inclines.

For gravimetrically determining the loading of material conveyed on a pan feeder, some of the rail track is modified so that it is supported by load cells. These cells float in contrast to the fixed-wheel supporting rails, creating a large weighing platform similar to a truck or rail scale. The pan speed is detected by a speed sensor just like a belt conveyor scale system. The weight detected by the load cells, and the belt speed, are both fed into a belt scale integrator that determines the flow rate and totalizes the conveyed material.

Pan feeder application	
Mining	Yes
Aggregate	Yes
Cement	Yes
Food	No
Chemical, fertilizer (dry)	No
Grain	No
Steel and coal	Yes
Water/wastewater	No

Rotary airlock feeders

Rotary airlock feeders, also known as rotary feeders, rotary valves, star feeders, vane feeders, or any combination thereof, are designed to convey material by gravity from a bin, or chute, and maintain an airlock between two parts of the process or system. This airlock prevents easily fluidized material from flushing uncontrolled through a system.

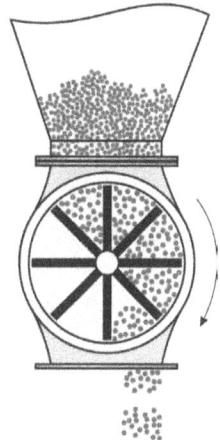

17

This seal combines with the positive conveying action to make these feeders ideal for introducing powder and granular materials to weighbelt feeders or solids flowmeters, thereby creating a process feed rate control system. The rotary airlock feeder is a good pre-feeder to a weighing device, and the more vanes the better so that its discharge flow has a high pulsation frequency so small pockets of material maintain a smooth flow pattern.

Rotary airlock feeder application	
Mining	Yes
Aggregate	Possible. Not common
Cement	Yes
Food	Yes
Chemical, fertilizer (dry)	Yes
Grain	Yes
Steel and coal	Possible. Not common
Water/wastewater	Possible. Not common

Aerated gravity conveyors

Aerated gravity conveyors (also known as aeration conveyors, fluidized gravity conveyors, air gravity fluidized conveyor, and Airslides®[2]) are designed for the enclosed transportation of fine powders and granules. They keep the material in a constant state of aeration allowing the aerated materials to flow on a cushion of air.

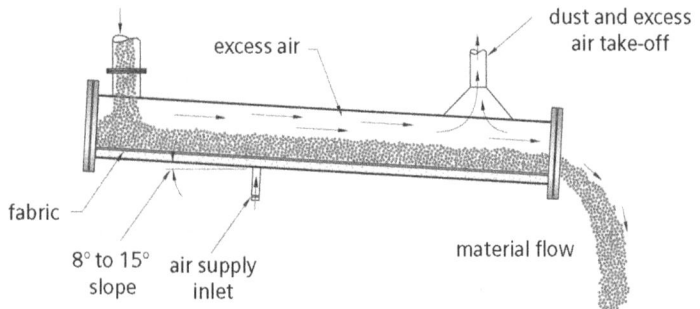

_____

[2] Airslide is a registered trademark of Fuller F.L. Smidth.

The flow of material can be volumetrically modulated with any of the following:

- vertical V-ball valves
- horizontal V-notch valves
- left unmodulated with slide gates serving as ON/OFF control devices

To gravimetrically weigh the material accurately, the material must be de-aerated before it can be sheared from a hopper by a weigh-belt feeder. Setups for proper de-aeration include the following:

- install a rotary airlock feeder as an interface between the bin and the weighbelt feeder
- use a solids flowmeter, making sure that the air flow does not affect the flowmeter's readings

Aerated gravity conveyer application	
Mining	Yes
Aggregate	No
Cement	Yes
Food	No
Chemical, fertilizer (dry)	Yes
Grain	Yes
Steel and coal	Yes
Water/wastewater	No

Gates, valves, and shutoffs

There are hundreds of different types, sizes, and styles of gates and valves, ranging from standard designs to those custom-engineered for very specific applications. The design principle is simple – an obstruction is inserted into the gravity flow of material, providing either ON/OFF control, or the modulation of flow, through the partial insertion of the barrier or by using diverter gates.

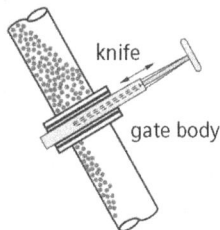

Some knife gate valves have v-notch or diamond notch cross sectional designs for better linearity of throughput relative to a

19

percentage of the open position so they can operate as modulating valves. Gates, valves, and shutoffs can be used with weighbelt feeders or solids flowmeters in the following applications:

- start and stop the material flow to the weighing device or chute
- set and adjust flow rates
- act as a modulating valve
- stop the flow of material for maintenance or calibration of dynamic weighing devices

Gates, valves, and shutoffs application	
Mining	Yes
Aggregate	Yes
Cement	Yes
Food	Yes
Chemical, fertilizer (dry)	Yes
Grain	Yes
Steel and coal	Yes
Water/wastewater	Yes

Rotating table feeders

Rotating table feeders provide the enclosed feeding of materials from a bin. This feeder controls powders, granules, or medium coarse materials by moving the material with rotating spokes or vanes. A cut-out in the non-rotating base allows material to fall out.

The flow of material is volumetrically modulated by varying the rotation speed of the spokes. To gravimetrically weigh the material, the discharged material can be introduced to a weighbelt feeder. The spoke rotation speed as well as the weighbelt speed can be varied to meet a set point.

Possible gravimetric solutions other than using a weighbelt:

- use a solids flowmeter, providing that the material size is not too great
- support the whole feeder on load cells and do loss-in-weight calculations to determine the feed rate

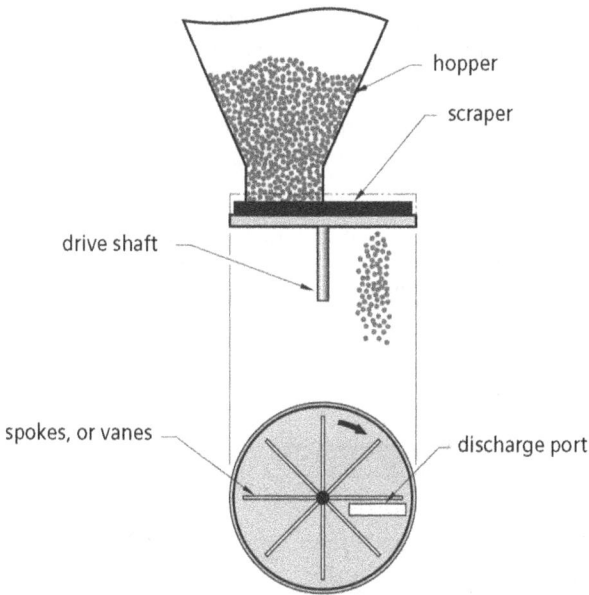

Rotating table feed application	
Mining	Yes
Aggregate	No
Cement	Yes
Food	Yes
Chemical, fertilizer (dry)	Yes
Grain	No
Steel and coal	Possible. Not common
Water/wastewater	Yes

Drag chain conveyors

Drag chain conveyors generally move material at a constant and slow speed, either horizontally or on an upward slope with the help of some design enhancements. Generally not used to modulate the flow rate, they can be supplemented with a weighbelt feeder or a solids flowmeter should the flow rate need to be monitored.

The typical heavy pulsating flow pattern from the drag chain requires modifications to ensure a smooth flow of material to the weighing device. There are several choices:

- use special inserts in the discharge of the drag conveyor
- add a surge bin between the drag conveyor and the dynamic weighing device

Monitor the surge bin and consistent shearing conditions with load cells or an ultrasonic sensor, or the new SITRANS LR560 radar transmitter, to ensure the following get consistent flow rates:

- screw feeder
- weighfeeder
- gravity flow conditions for a solids flowmeter

Drag chain conveyer application	
Mining	Yes
Aggregate	Possible. Not common
Cement	Yes
Food	No
Chemical, fertilizer (dry)	Possible. Not common
Grain	Yes
Steel and coal	Yes
Water/wastewater	Possible. Not common

Bucket elevators

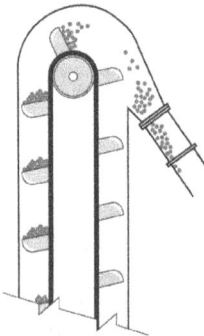

Bucket elevators come in varying sizes and types and are engineered specifically for elevating material vertically. Using buckets made of a variety of materials, including plastics, thermoplastics, mild steel, and stainless steel, they convey material upward at a constant rate. While chain driven versions run slowly and belt driven versions move more quickly, the elevators are not generally used to modulate the material flow rate.

To determine the material transfer rate, a weighbelt feeder or a solids flowmeter can measure the discharge flow. The material flow must run smoothly and evenly through the weighing device with reduced velocity in order to limit abrasion.

Reduce the velocity by placing special inserts in the device discharge. Alternatively, a surge bin is placed between the bucket elevator and the weighing device. The surge bin level should be monitored, either by load cells or by a non-contacting level technology such as an ultrasonic sensor or the new SITRANS LR560 radar transmitter.

Bucket elevator application	
Mining	Yes
Aggregate	Possible. Not common
Cement	Yes
Food	Possible. Not common
Chemical, fertilizer (dry)	Yes
Grain	Yes
Steel and coal	Yes
Water/wastewater	No

Vibratory feeders

Vibratory feeders move or extract material that is not free flowing or tends to bridge or rat-hole in a bin.[3] These feeders move material laterally by vibrating a sloped bed. By varying the amplitude and the frequency of the vibration, granular and larger materials move at selectable volumetric rates.

To obtain gravimetric measurement, the unit can feed onto any of the following:

- a weighbelt feeder
- a solids flowmeter
- a belt conveyor using a belt scale

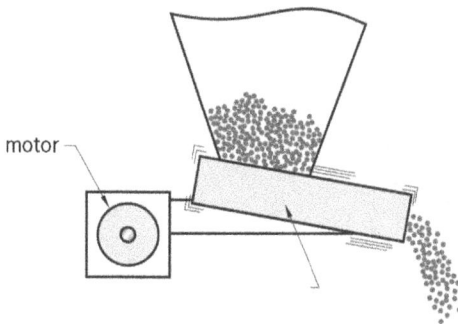

motor

---

[3] When material sticks together creating empty pockets under the surface.

Vibratory feeder application	
Mining	Yes
Aggregate	Yes
Cement	Yes
Food	Yes
Chemical, fertilizer (dry)	Yes
Grain	Yes
Steel and coal	Yes
Water/wastewater	Yes

Summary

The mechanism used for moving the material is determined by the material type, the direction of the transport, the length of the transport, and the accuracy requirements. Quite often, the only requirement is the distance of transport; however, when the weight of the transported material and its accuracy become factors, a weighing device needs to be incorporated. The next chapter discusses numerous common weighing devices available in the market place, including Siemens Milltronics belt scales and solids flowmeters.

Chapter 3

Dynamic Weighing Devices

You load sixteen tons, what do you get?[1]

Transporting dry bulk solid materials uses a variety of methods to get material from one point to another. Whether it be for storage, stockpiling, process feeding, packaging, blending, receiving, production monitoring, or loadout, the material often needs to be weighed and have its flow rate controlled. This requirement is fulfilled by a dynamic weighing device that can calculate the weight of material on the move.

These weighing devices improve plant efficiencies, automate plants and processes, manage inventory, control quality, and can legally be used for trade measurements.

A quick application reference chart follows each section outlining typical applications where the device is used.

Topics

There are five principal devices/systems for dynamic weighing of dry bulk solid materials while in motion.

- Belt scales
- Weighfeeders and weighbelt conveyors
- Solids flowmeters
- Loss in weight
- Rotary weigh feeder

---

[1] Ford, Tennessee Ernie. "Sixteen Tons." 1955. First recorded in 1946 by American country singer Merle Travis.

Belt scales

Belt scales (aka: weightometers or belt weighers) have been used for over 100 years. Applied to belt conveyors and belt feeders, they monitor the feed rate at which material is being conveyed (i.e. tons per hour) and keep a running total of the product being conveyed (totalization). Initially fully mechanical,[2] these devices now use mechanical to electrical transducers to sense a conveyor load, as well as rotation sensors to detect the belt speed.

The load and speed measurements are processed by a belt scale integrator which calculates rate of flow, belt loading, belt speed, and material totalization. Belt scales come in a variety of constructs, shapes, sizes, and ranges to suit most belt conveying applications.[3]

Belt conveyor with belt scale: pre-feeding device			
Belt conveyor	Yes	Vibratory feeder	Yes
Belt feeder	Yes	Apron feeder	Yes
Screw conveyor	Yes	Pan conveyor	Yes
Screw feeder	Yes	Rotating table	Yes
Bucket elevator	Yes	Slide gate valve	Yes
Drag chain conveyor	Yes	Bin/silo/hopper	No
Aerated gravity conveyor	Possible: difficult to de-aerate product	Rotary airlock	Yes

Weighfeeders

A weighfeeder is a belt feeder designed specifically for weighing and feeding. Commonly used to extract material from a holding bin or hopper at a set rate, it regulates the material using a shearing action. The rate is derived and calculated by an electronic belt scale integrator using the output of the belt scale load sensing transducer combined with a belt speed pulse stream.

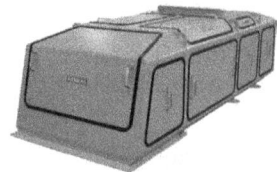

The calculated rate is compared to a process set point and the belt speed of the weighfeeder is continually adjusted, allowing the feed

---

[2] See *Chapter Seven* for more information on mechanical integrators.
[3] This book focuses mainly on belt scales, but also discusses other weighing systems to set the context.

rate to match the required feed rate set point. Weighfeeders are generally custom designed to suit specific application dimensions and material characteristics for industry and environmental conditions.

Weighfeeders and weighbelts come either with flat rollers or with slightly troughed idlers supporting the belt. In lower feed rate applications, a flat low friction frame constructed from metal, which the belt slides on, is often supplied. This floating, flat, low-friction frame is supported by load cells and is referred to as a weigh deck.

Weighfeeders: pre-feeding device			
Belt conveyor	Yes	Vibratory feeder	Yes
Belt feeder	Yes	Apron feeder	Yes
Screw conveyor	Yes	Pan conveyor	Yes
Screw feeder	Yes	Rotating table	Yes
Bucket elevator	Yes	Slide gate valve	Yes
Drag chain conveyor	Yes	Bin/silo/hopper	Yes, most common
Aerated gravity conveyor	Possible: difficult to de-aerate product	Rotary airlock	Yes

Weighbelt conveyors

Similar to a weighfeeder, a weighbelt conveyor is designed for constant speed operation. A pre-feeding device such as a vibratory feeder, screw conveyor, or rotary airlock feeder sets the volume of material that goes onto the belt.

The belt scale and integrator provide the signal to control the speed of the pre-feed device and to regulate the material flow rate to the weighfeeder. Although a weighbelt is generally a constant belt speed device, the weighbelt conveyor speed can be varied proportionally with the flow rate from the pre-feed device. This feature allows it to maintain a more consistent bed depth, thus establishing greater accuracy.

Weighfeeders and weighbelt conveyors are typically short in length and are designed to achieve the best weighing accuracy while providing feeding, conveying, feed rate modulation, dosing, and batching functions.

Weighfeeders and weighbelts come either with flat rollers or with slightly troughed idlers supporting the belt. In lower feed rate applications, a flat low friction frame constructed from metal, which the belt slides on, is often supplied. This floating, flat, low-friction frame is supported by load cells and is referred to as a weigh deck.

Weighbelt conveyors: pre-feeding devices			
Belt conveyor	Yes	Vibratory feeder	Yes
Belt feeder	Yes	Apron feeder	Yes
Screw conveyor	Yes	Pan conveyor	Yes
Screw feeder	Yes	Rotating table	Yes
Bucket elevator	Yes	Slide gate valve	Yes
Drag chain conveyor	Yes	Bin/silo/hopper	No
Aerated gravity conveyor	Possible: not recommended	Rotary airlock	Yes

Solids flowmeters

Solids flowmeters measure the flow rate of dry bulk solid, powdered, or granular materials and are widely used in chemical, plastics, cement, aggregate, mining, and food and pharmaceutical industries. They are employed where the free flowing material is fed by gravity through pipes, ducts, or chutes through various stages of processing, storage, blending, receiving, or loadout.

There are three significant solids flowmeters designs.

1. Impact
2. Centripetal or momentum force
3. Coriolis force

Impact

Measures the horizontal force of material impact on a sensing plate with the vertical forces negated by the design of the sensing plate suspension mechanism.

Two ways to measure the impact force:

- strain gauge load cell: a transducer that converts horizontal force to an electrical signal creates a signal proportional to the flow of material stream hitting the sensing plate.

- Linear Variable Differential Transformer (LVDT): measures the horizontal deflection caused by the impact forces. This design often uses a viscous fluid damper to act as a shock absorber, useful in applications with pulsating flow or occasional flow surges. Some LVDT models use frictionless pivots to reduce friction, eliminate hysteresis, and improve sensitivity. This design also make it easier to measure flow rates as low as 100 kilos per hour full scale.

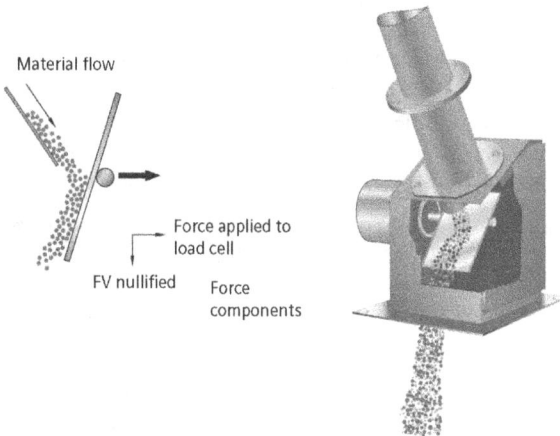

Material flow

Force applied to load cell

FV nullified

Force components

Some manufacturers, like Siemens, use the mechanical design to nullify the vertical component of the impact force, leaving only the horizontal component to be measured. As material builds up on the sensing plate, thereby increasing the mass of the sensing plate, this design allows the position of the sensing plate to remain unchanged, thus having no effect on the zero calibration of the system.

I was standing on my bathroom scale one morning when my wife walked by and saw me looking down at the scale with a frustrated expression. I then sucked in my stomach, looked down again, appeared relieved, and said, " Ah, that's better."

My wife stopped and said: "Sucking in your gut doesn't make any difference in your weight."

"I know," I replied, "but at least now I can see the numbers!"

Centripetal or momentum force

This flowmeter measures the centripetal force created as the flow of material changes over a curved sensing plate or measuring chute instead of measuring direct impact forces.

- The curved sensing plate is supported by specifically designed mechanical structures that connect to load cells to measure either the horizontal and vertical components of force, or just one force generated by the material accelerating through the curve. This measured centripetal acceleration force is then interpreted as an instantaneous flow rate of material.
- Changes in bulk density do not affect accuracy; however, material buildup equating to added mass to the sensing plate will cause a positive shift in the static zero. This solids flowmeter does not measure as low a flow rate as an impact flowmeter.

Impact and centripetal solids flowmeters: pre-feeding devices			
Belt conveyor	Yes (not common)	Vibratory feeder	Yes
Belt feeder	Yes (not common)	Apron feeder	Yes
Screw conveyor	Yes	Pan conveyor	Yes
Screw feeder	Yes	Rotating table	Yes
Bucket elevator	Yes	Slide gate valve	Yes
Drag chain conveyor	Yes	Bin/silo/hopper	Yes
Aerated gravity conveyor	Yes	Rotary airlock	Yes

NOTE: Best suited for particles not larger than 25 millimeters (1 inch). Some applications may be limited due to particle size.

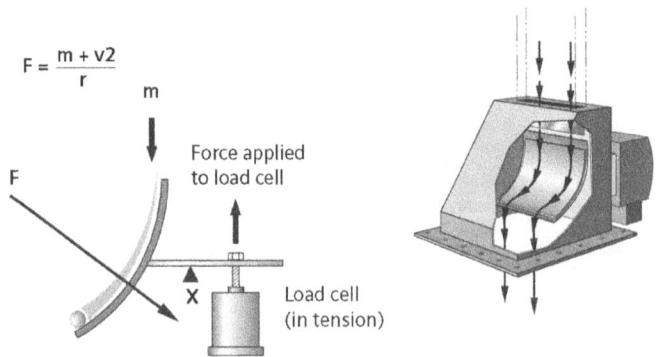

$$F = \frac{m + v2}{r}$$

m

Force applied to load cell

F

X

Load cell (in tension)

Coriolis force

This flowmeter uses the Coriolis effect as a measuring principle. The effect is created when bulk solids make contact with a constant speed and vaned measuring wheel.

As the bulk solid particles make contact with the vanes of the rotating measuring wheel they move outwards in a radial path creating the Coriolis force. This force on the measuring wheel creates a change in torque on the motor driving the measuring wheel. The torque change is then measured and converted to an electrical signal by either an accelerometer or a torque transducer (load cell) attached to the frame of the motor. It's possible to add a tachometer and motor speed controller to ensure constant impeller speed, further enhancing accuracy.

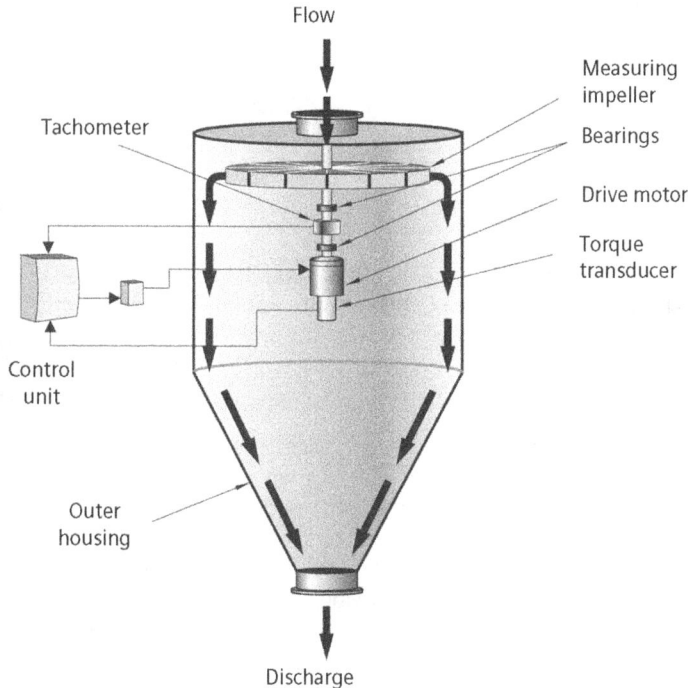

Flow

Measuring impeller

Tachometer

Bearings

Drive motor

Torque transducer

Control unit

Outer housing

Discharge

The electrical signal is converted into an instantaneous flow rate of material, and accuracy is not affected by changing bulk density or different grain size. This device is best suited to powders, especially those that will not stick and eventually bind the mechanisms. Accuracy achieved by this type of device can be quite impressive.

31

Coriolis solids flowmeter: pre-feeding devices			
Belt conveyor	Possible (not common)	Vibratory feeder	Yes
Belt feeder	Possible (not common)	Apron feeder	Possible (not common)
Screw conveyor	Yes	Pan conveyor	Possible (not common)
Screw feeder	Yes	Rotating table	Yes
Bucket elevator	Possible (not common)	Slide gate valve	Yes
Drag chain conveyor	Possible (not common)	Bin/silo/hopper	Yes
Aerated gravity conveyor	Yes	Rotary airlock	Yes

NOTE: Best suited for powdery materials. Applications may be limited because of particle size.

Loss-in-weight feeders

Loss-in-weight (LIW) systems are commonly used with lower flow rate of bulk solids, as well as the control of liquids by weight. Usually a smaller holding hopper and a mechanical volumetric feeder are supported entirely by load cells which monitor the loss of weight in the hopper as material is volumetrically removed.

The speed of the volumetric feeder is controlled to modulate the flow rate of material. The feeding device, often a screw auger or a rotary feeder, needs to fill consistently and uniformly, thereby delivering a linear volume of material proportional to the speed of the feeding device. An electronic indicator monitors the load cell signal and speed of modulating device, deriving a flow rate over time assuming constant bulk density and a well-designed modulating device.

NOTE: Pre-feed chute work to the hopper needs to be isolated as much as possible from the hopper on load cells. Also any chute work after the suspended discharge device needs to be kept away from the hopper. If not, any discharge affects the accuracy of the weighing process.

Charging (filling) stream

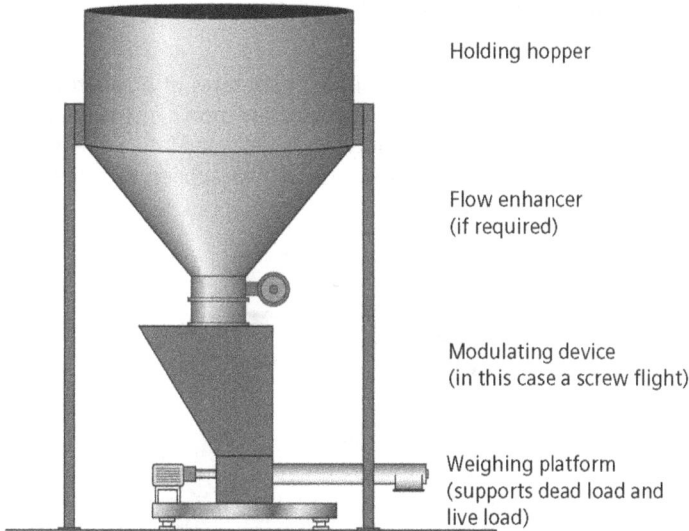

Holding hopper

Flow enhancer
(if required)

Modulating device
(in this case a screw flight)

Weighing platform
(supports dead load and
live load)

Weighing accuracy is generally very good with LIW systems, especially if the hopper and feeder can be well isolated from other process components. However, accuracy can be affected at the time of re-filling the hopper in a continuous running situation. During the hopper filling time, the measurement of the loss of weight and subsequent calculated rate of flow is hampered by the addition of weight to the hopper. As a result, the feeding device needs to be set to a constant speed and operated volumetrically during the hopper fill time.

For best average operational accuracy, the hopper filling time needs to be kept at a minimum. This often necessitates a "charge" bin where a predetermined amount of material is accumulated prior to the filling cycle and can be added quickly during the volumetric mode. After the filling is complete and the load cell signals stabilize, the LIW control processing can begin once again.

Rotor weighfeeder

The rotor (or rotary) weighfeeder is a like a horizontal rotary valve or table feeder adapted for gravimetric weighing. The momentary load on the table is measured by a load cell(s) and the table's rotational speed is adjusted to maintain a constant feed rate.

The material inlet is almost always a full chute from a silo, or surge bin, although some mechanical feeders may be suitable as pre-feeding devices.

This device is often used to maintain flow rates of powdered and granular materials (e.g. pulverized coal, raw meal cement) through the rotor weighfeeder and then typically delivered into a pneumatic conveying system. Or, the discharge can be applied to chutes for controlled truck and rail car loading.

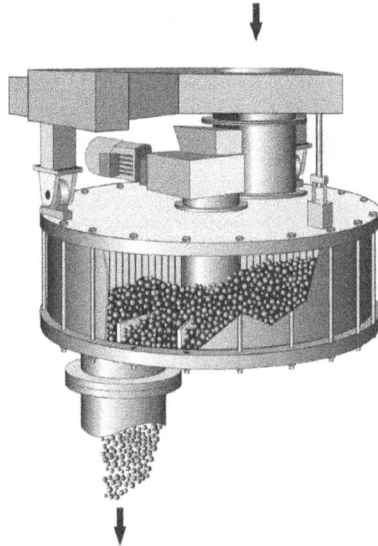

Summary

Measuring moving material effectively depends on a great number of factors, including volume, speeds, mass, size, and direction. The measurement device chosen needs to take all these factors into account as well as the requirements of the application – these include accuracy requirements, cost restrictions, and location needs. The bulk of applications, however, can be weighed using a belt scale. The rest of this book will focus primarily on belt scale systems: how they work, the components, and the variations.

Chapter 4

Belt Scale Designs

Put the load right on me.[1]

The belt scale is at the center of dynamic weighing, and Siemens has a broad range of single and multi-idler configurations available for process and load-out control. Whether enclosed in a weighfeeder or on their own, Siemens Milltronics belt scales provide continuous in-line weighing of a variety of products in primary and secondary industries from extraction (in mines, quarries, and pits) to power generation, iron and steel, food processing, and chemicals.

> **NOTE:** A terminology clarification is required here as belt scale, weighbridge, and belt weigher are often used interchangeably, even though they each have a specific role:
>
> **weighbridge:** the actual instrument supporting the weighing idler(s)*
>
> **belt scale or belt weigher:** the system containing the weighbridge
>
> *Please note that design and application references to 'idler' also apply to a multi-idler design.

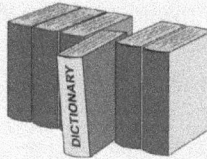

This chapter looks at how the belt scale weighbridge is designed and used most effectively. Both single and multi-idler belt scales come in a variety of configurations, and their performance quality depends on both their design and application.[2]

Topics

- Weighbridge design
- Design objectives
- Design considerations
- Current belt scale designs

---

[1] The Band, "The Weight." *Music from the Big Pink*, 1968.
[2] For more on applications, please see *Chapter Nine*.

- Common full construction single-idler weighbridge designs and comparisons
- Single idler: full construction versus modular design
- Installing a modular single-idler scale
- Common multi-idler weighbridge designs
- Multi-idler: full length versus multiple single idlers
- The MSI – advanced weighbridge designed for exceptional accuracy

Weighbridge design

A belt scale weighbridge is the mechanical structure supporting a conveyor idler and thus the moving belt. The bridge supports the weighing idler within the belt conveyor, separate from the other carrying idlers which are fixed and just roll. The weighbridge is the mechanical transfer point, permitting the measurement of the weight of the conveyed material.

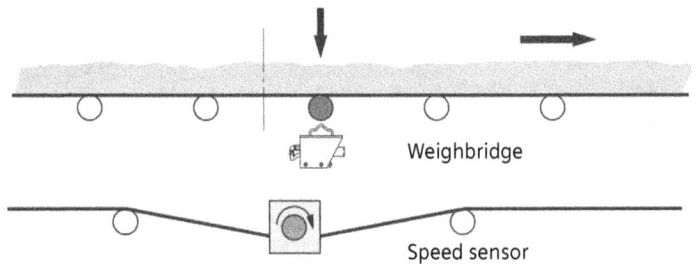

Weighbridge

Speed sensor

The weighbridge has four basic components:

- Dynamic frame: supports the weighing idler
- Static frame: rigidly supported by the belt conveyor structural frame
- Mechanics: transfers the live load which is the weight of material conveyed by the belt as well as the tare load (or dead load) to a mechanical-to-electrical transducer or to a load cell:
    - the weight of the dynamic frame
    - weight of the idler
    - weight of the belt section in the scale area
- Mechanical-to-electrical transducer or load cell: produces an electrical signal proportional to the amount of tare loading and vertical forces applied to the weighing roller by the conveyed material load on the belt.

Weighing idler, or roller.
Supported by weighbridge
and floats with respect to
adjacent idlers.

Dynamic frame
(rigidly supports
weighing idler)

Mechanics with
load cell (heart of
the design).

Counter weight

Weighbridge static structure frame,
rigidly supported by conveyor structure.

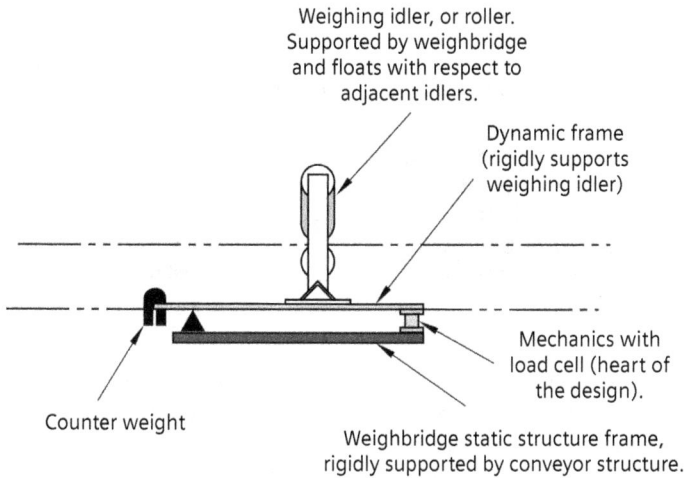

Belt scale weighbridge design

A successful belt scale system must provide accurate, reliable, long-term service so that it has a reasonably short-term payback, making cost of ownership an attractive proposal. Consider the following as part of the design process in order to get the best out of the installation.

Accuracy – the primary objective which is achieved by the following:

- limiting vertical deflection because belt loading generates vertical forces through the belt to the weighing idler
- ignoring horizontal forces resulting from the rolling load of the belt and idler friction
- ensuring high resolution mechanical-to-electrical transduction from vertical forces generated by conveyor loading
- reacting quickly to changes in the vertical forces that allow readings to be almost instantaneous, avoiding the reaction time delay of the weighbridge mechanics

Ease of installation – avoid excessive structural weight and size of the weighbridge design. Installation should not require excess resources or heavy lifting equipment. Ideally, the weighbridge should be installable by no more than two people.

37

Ease of calibration – electronic calibration of the system is preferred as it avoids using reference weights or calibration test chains. The current trend, however, is still to use static test weight or test chains because they apply gravitational forces as calibration references. Therefore, a weighbridge design allowing the addition of static weights to the dynamic load frame, either manually or by mechanical device, is required.

> NOTE: Material tests are required to validate calibration and to recognize any errors caused by the belt's movement and ability to support the loading of material.

Ease of maintenance – limit surfaces for material to gather on the dynamic structure because the extra weight from the buildup will offset the empty belt zero reading. The static portion of the scale, consisting of the fixed structure supported by the conveyor support frame plus the load cell(s), should not impede the movement of the dynamic structure of the scale, even if buildup occurs.

Make sure the load transducer is protected so that cleaning with pressurized air or water will not cause damage. Mechanical stops should limit excess forces and the resulting deflection to the dynamic beam, thus protecting the load cells.

> ⚠ CAUTION: Excess forces can come from conveyor over loading and from accidental shock loads. Avoid the following:
> • walking on the weighing idlers
> • dropping equipment on weighing idlers
> • temporarily storing equipment on the weighing idlers

Long term reliability – replacing the weighing system components means more than additional costs; it leads to a loss from production down time. Reliable performance also reduces costs by minimizing routine maintenance and component replacement.

Weighbridge design considerations

To meet the system's objective of providing long term and accurate service, the weighbridge design phase needs to take the following into account.

Strong support to avoid deflection:

- integrity of the structural metal must be sufficient for extreme conditions
- the weighing idler must be supported without excessive deflection occurring in order to maintain its alignment with the adjacent carrying idlers. Excess vertical deflection of the weighing idler causes it to move away from the conveyor belt, translating into non-linear weighing over the conveyor loading range.
- excess deflection of the load cell may combine with conveyor conditions due to harmonic vibration to cause the weighing idler to oscillate
- strain gauge load cells need to be selected for limited deflection
- if an LVDT is the transducer, make sure the deflection necessary for operation of the LVDT does not cause excessive vertical movement of the weighing idler
- the weighing idler must be firmly secured to avoid lateral movement of the idler across the conveyor

Likely sources of weighing idler deflection:

- deflection of weighbridge dynamic structure supporting the weighing idler
- normal deflection of load cell under load
- deflection of static structure of the weighbridge
- deflection of the conveyor stringers, or conveyor stringer supports under loaded conditions (often beyond the designer's control)

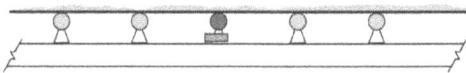

Under light loading, idler alignment is maintained.

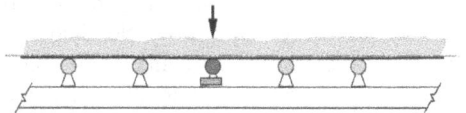

Under heavy loading, accumulated vertical deflection can cause the weigh idler to move away from the conveyor belt line.

Design must be in harmony with conveyor to avoid oscillation of weighing idlers.

Only the vertical forces should be measured, with the horizontal forces ignored.

Vertical deflection must be minimal.

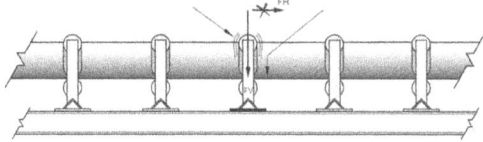

Structural integrity and transducer deflection are two very important aspects of weighbridge design.

> **NOTE:** The Siemens Milltronics belt scale application program calculates any risks of weighing oscillations by using algorithms developed by Siemens Milltronics R & D in consultation with the Queen's University* mechanical engineering department.
>
> *Located in the charming city of Kingston, Ontario, Canada.

The application of the load to the transducer or load cell – direct application is best. However, many belt scale manufacturers use semi-indirect, or indirect, application methods using pivots and levers. The most effective design will apply both the dead load of the scale system (conveyor idler, belt, and scale structure) and the live load as they are applied directly to the load cells.

The weighbridge design and load cells should combine so that only the vertical forces related to conveyor loading are measured – not the horizontal forces that are the result of bearing drag and roller-to-belt friction.

Current weighbridge designs

Single-idler belt scales have dominated the weighing market in recent years as they meet the accuracy demands of blending systems and provide accountable performance. Coupled with the need for lower cost of ownership belt scales, the single-idler scale is a popular choice.

Numerous vendors offer full construction single-idler belt scale systems, many operating with the very effective dual-load cell design. Continual customer demands for even lower cost systems have led to modular belt scales supporting a single weighing idler with two weigh modules, one on each side of the conveyor. Neither a static beam, nor a dynamic beam are used, with only the weighing idler forming a cross conveyor structure.

The traditional multi-idler belt scale weighbridge remains available, but the single-idler designs have a much larger share of the business, especially when higher accuracies are required. Instead of using the large, cumbersome multi-idler weighbridges, two, three, or even four single-idler belt scales are installed in tandem, creating a highly accurate modular system.

There are three basic full construction, single-idler weighbridge designs:

- *basic pivoted design* (traditional) – the weighing idler is supported by a lever structure that balances on a pivot point applying forces to one or two load cells, depending on belt width.
- *single leaf spring as flexure bearing* – the design is between the basic pivoted design and the non-pivoting design. It applies the weighing idler directly to the load cell; it still has a flexure bearing for stability sake.
- *non-pivoting design* – the idler loading is applied directly to load cells.

This section discusses the following:

- the current design of single-idler belt scales and multi-idler scales
- a comparison of full construction single-idler belt scales to modular scales

1. Basic pivot

The pivot design is based on the lever,[3] one of the six simple machines found in basic mechanics. It can be used in both single-idler and multi-idler weighbridges. There are several variations on this basic design.

The diagram below shows the dynamic frame (lever) of the weighbridge rotating through a pivot point and applying the *dead load* (the shared weight of dynamic frame plus weighing idler plus conveying belt) and the *live load* (the conveyed material loading on the conveyor). A static weight can be used on the opposite side of the pivot to counter balance some of the weight of the dead load. There is usually a pivot point at each side of the conveyor, but there are some exceptions; and the style of pivots varies among manufacturers.

---

[3] The six classical simple machines were defined in the Renaissance: lever, wheel and axle, pulley, inclined plane, wedge, and screw.

Top view

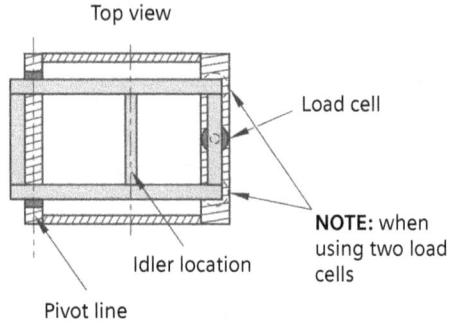

Load cell

NOTE: when
using two load
cells

Idler location

Pivot line

The most common designs currently in use are reviewed below:

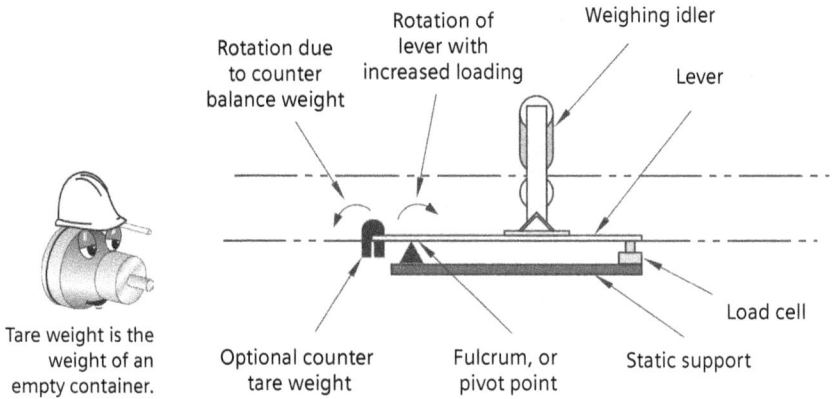

Rotation due
to counter
balance weight

Rotation of
lever with
increased loading

Weighing idler

Lever

Load cell

Tare weight is the
weight of an
empty container.

Optional counter
tare weight

Fulcrum, or
pivot point

Static support

Pivots, load cells, and check rods

Pivots: the original pivot points used in belt scales were *knife* edges supporting a lever that had hardened grooved surface points. These pivot points would wear and were easily contaminated, impeding movement and leading to inaccuracies.

Eventually, most suppliers changed to other pivot styles to reduce the maintenance requirements and establish the pivot location as close as possible to the belt line to reduce errors. *Roller bearings* and *needle bearings* are also used but will eventually be contaminated or wear to the point that hysteresis within the bearing affects repeatability and accuracy.

Bottom of belt — Pivot – load cell line

Knife edge pivot or stainless steel torque shaft — Load cell

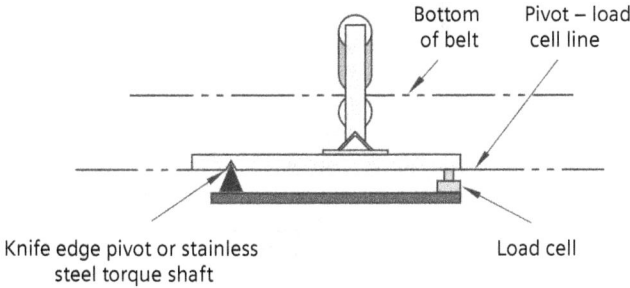

These pivot alternatives are available to compensate for these performance problems:

- *rubber trunnions*: improve performance but the rubber is affected by ambient temperature changes. It also hardens and becomes brittle with age. Both these aspects affect accuracy.
- *stainless steel shafts*: used the same way as a continuous torque shaft trunnion. They do not deteriorate with age like rubber, but its torsion properties do change with temperature.
- flexure bearings are an improvement over roller bearings and needle bearings and are affected by corrosion and temperature variation

A trunnion is a protrusion used as a mounting point.

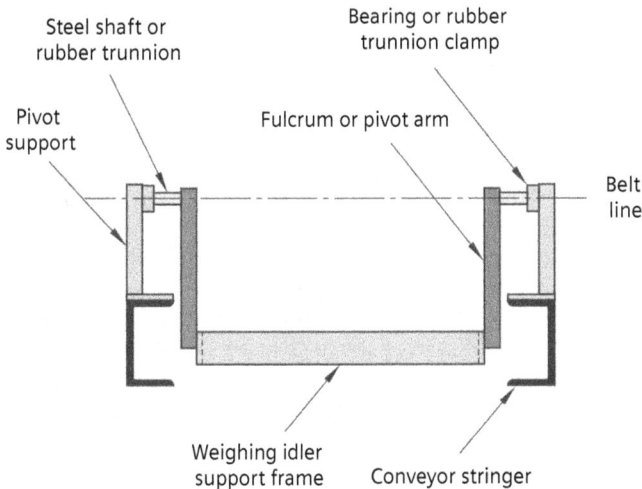

Steel shaft or rubber trunnion — Bearing or rubber trunnion clamp

Pivot support — Fulcrum or pivot arm — Belt line

Weighing idler support frame — Conveyor stringer

NOTE: Counter-torque can be applied to the stainless steel shaft, thereby creating a counter tare force and reducing the dead load applied to the load cell.

While visiting a mining company, I walked past their multi-idler scales. They were not Siemens scales and used bearings as their pivot points. As I had a closer look, I noticed a sledge hammer and a piece of wood beside each scale.

Seeing my puzzled look, the operator told me it was improvised quality control to correct weighing errors. Every few days, they would hold the wood to the bearing and smack it with a hammer. This jarred the bearings, providing short lived correction of hysteresis affecting the belt scale pivot points.

The MSI does not require a hammer!

Load cells: the compression load cell is still used, even though it needs a very stable relationship with the pivot point. Wearing pivots can change that location and cause changes in accuracy. An improvement over the compression load cell is the S load cell applied in tension through a stainless steel wire cable. The S cell is far less affected by side forces caused by poor alignment of the applied load and the load cell.

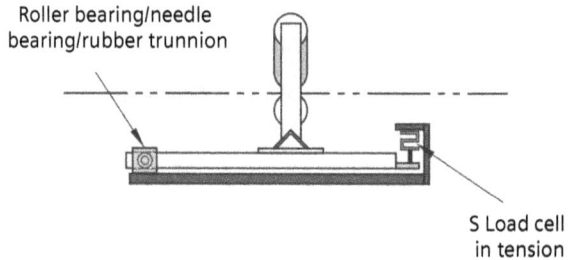

Roller bearing/needle
bearing/rubber trunnion

S Load cell
in tension

Check rods: the long thin bolts designed to eliminate horizontal movement of the weighbridge.[4] Their length and size allows them to act as a flexure bearing, unfortunately causing the vertical weighing forces to be impeded as they are transferred to the load cells. That is similar to stabilizing yourself on the bathroom sink as you weigh yourself on a bathroom scale – readings are thus inaccurate and non-repeatable.

A flexure bearing is a bearing which allows motion by bending a load element.

---

[4] Some weighbridge designs use as many as four check rods.

Dynamic   Check
weighbridge   rod
frame

Check rod sized to check   Static frame
horizontal movement and
flexible enough to allow
impeded vertical movement.

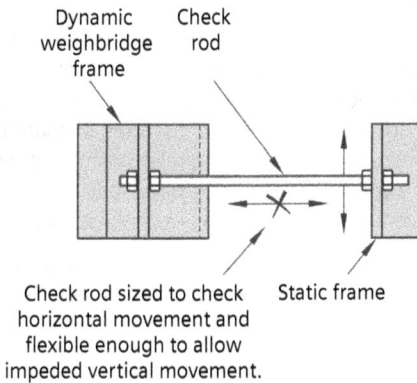

Limitations of the traditional pivoted belt scale

The design is limited. For optimum performance the pivot point should be as close to the belt line as possible, and the pivot-load cell line should be parallel to the belt line. This ideal placement eliminates the effect of horizontal forces on the load cells created by the rolling load. However, given the generally cramped nature of the applications, this placement is often not possible or easily done.

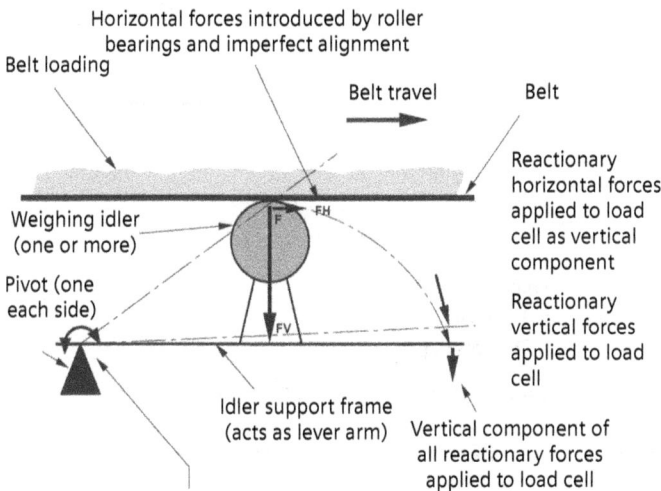

Horizontal forces introduced by roller
bearings and imperfect alignment

Belt loading

Belt travel   Belt

Weighing idler
(one or more)

Reactionary
horizontal forces
applied to load
cell as vertical
component

Pivot (one
each side)

Reactionary
vertical forces
applied to load
cell

Idler support frame
(acts as lever arm)   Vertical component of
all reactionary forces
applied to load cell

All forces must rotate through pivot creating three issues:

• Time to rotate clockwise and counter clockwise
• Frictional forces introduced by the pivot
• Counter forces of pivot

45

Reaction time

Reaction time is a crucial component of accuracy because the sooner the load reading is transferred to the integrator, the more relevant it is. Furthermore, if the reaction time of a weighbridge is too slow, it may not measure minute changes in loading, especially with faster conveyor belts. Thus the relationship between conveyor belt speed and the distance of the idler spacing in the scale area must be as favorable as possible to ensure best accuracy.

For a complete transfer of belt loading to the weighbridge load cells, specific proven retention time (in seconds) is required. We suggest that the reaction time of the MSI is less than half of the retention time required for pivoted scales.

Retention time formula	
Number of idlers × idler spacing (feet)/ belt speed (fpm) / (60 sec/min)	
OR	= seconds
Number of idlers × idler spacing (m)/ belt speed (m/s)	

Pivoted belt scale

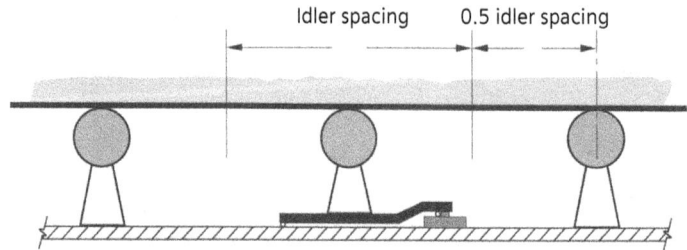

Idler spacing   0.5 idler spacing

Number of weighing idlers: 1
Idler spacing: 1 meter
Belt speed: 1.75 meters per second

Retention time = 1 × 1 m / 1.75 m/s = 0.57 seconds

> **NOTE:** Minimum scale retention time to meet the specification of a pivoted weighbridge is one second.

Belt scale with direct application of forces to the load cell

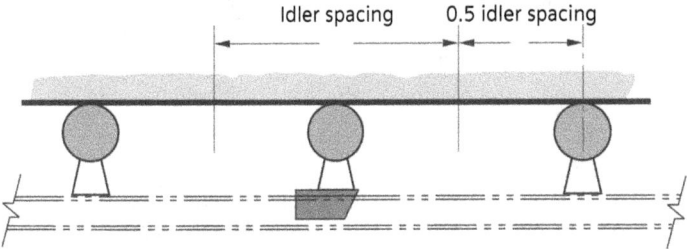

Idler spacing   0.5 idler spacing

Retention time	
1 × 4 ft. / (400 fpm / 60 sec/min)	= 0.6 seconds

NOTE: Minimum scale retention time specification was 0.5 seconds when the MSI belt scale was first introduced to the market. That has dropped a 0.4 second minimum and can go as low as 0.25 seconds in consistent loading applications. High accuracy requirements are now expected, even at greater belt speeds.

2. Leaf spring as flexure bearing

In this design, the leaf spring is quite wide and thick and serves as a flexure bearing. This design mounts a weighing idler to a cantilever beam load cell.

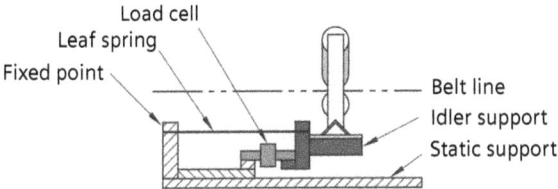

Load cell
Leaf spring
Fixed point
Belt line
Idler support
Static support

Benefits:
- fewer problems with a rolling load because the leaf spring is close to the belt line
- reaction time is faster than the traditional pivoted belt scale[5] since the weighing idler is mounted directly to load cell. The leaf spring slows reaction down somewhat.

[5] Please note that the leaf spring reaction time is not as fast as the direct-mount Siemens MSI scale.

47

Drawbacks:
- temperature changes affect the leaf spring's bending repeatability
- the leaf spring can be subject to corrosion and wear
- questionable stability, since side forces applied to the weighing idler can cause twisting of the leaf spring and the load cell

3. Direct application to dual load cells

The best single-idler belt scale design has the weighing idler directly supported by two single-point, parallelogram load cells.[6] This load cell style is usually the variation of a parallelogram read by strain gauges and is calibrated so the forces are read the same where ever they are applied on a predefined platform size.

Platform size can vary from 12 to 24 square inches (30 to 60 centimeters), depending on the load cell capacity.

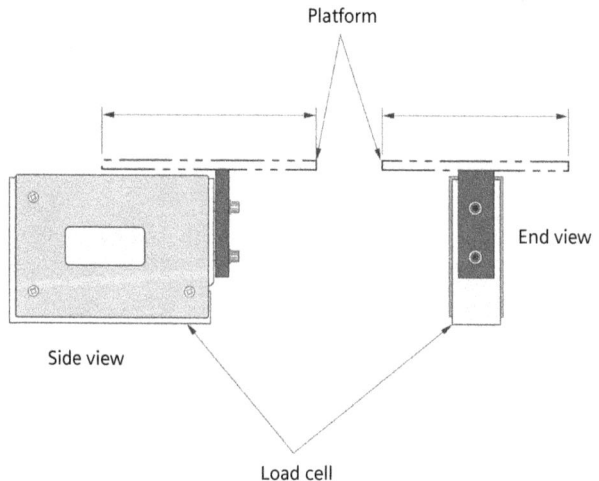

Platform

End view

Side view

Load cell

Two actively balanced load cells, one at each side of the conveyor, create a large and unique weighing platform that reacts to forces with the same reading, regardless of where the load is applied to the weighing idler.

Benefits
- rolling loads are no problem because there are no levers or pivots
- fast reaction time because the weighing idler is mounted directly on the two load cells, thus it weighs accurately even at high belt speeds

---

[6] Single point refers to an individual load cell rather than combined with several others.

Most load cells used in this configuration are dual-beam platform type; however, the triple-beam platform provides an even greater performance.[7]

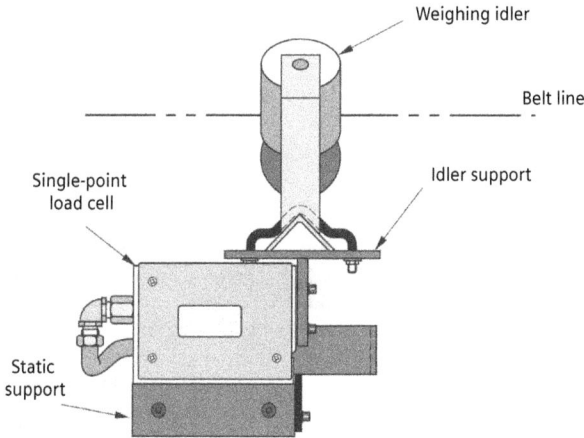

Common full construction single-idler weigh-bridge designs

Hanging modules using single-point load cells[8]	SIPWB[9] canister style load cell in compression
• the two modules are hung loosely from two pipes serving as structural cross members • these pipes are clamped to the conveyor stringer with U-bolts at each of the four stringer interface points • at each of these four points, two bolts with locking nuts act as jack-bolts to allow final idler alignment	See *Basic Pivot* design description *Chapter 4*, page 4 for more information.

---

[7] For more on the triple-beam platform, see *Chapter Five*.
[8] While the direct application of forces to the load cells is effective and makes for quick reaction time, the design can lead to mechanical hysteresis and non-linear weighing. Mechanical hysteresis is the inability of the instrument to provide repeatable results for given loading conditions. Non-linear weighing is when the loading is accurately reported at light conditions, but over reported at heavy loading conditions or vice versa. The stability is affected by the flexing of the structural cross members pipes under loading conditions, as well as the shifting of idler template alignment and horizontal alignment as a result of material loading.
[9] Single-idler pivoted weighbridge.

49

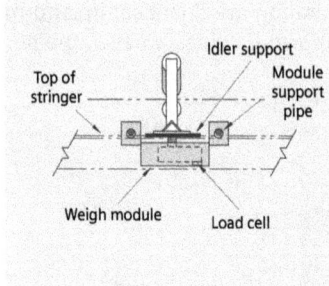

Performance rating: POOR

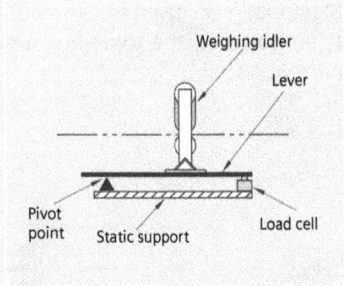

Performance rating: POOR

SIPWB S beam load cell in tension	SIPWB flexure bearing with idler applied at load cell
• pivoted suspension provides good stability • slow reaction time as the lever rotates through the pivots • pivots responding to wear and temperature variation may also contribute to mechanical hysteresis • S load cell in tension is an improvement over canister type in compression	• wide leaf spring acts as a flexure bearing to help with stability • slower reaction time as the flexure bearing rotates and bends with loading • temperature variation and wear on leaf spring can contribute to mechanical hysteresis • load cell is not designed for platform use

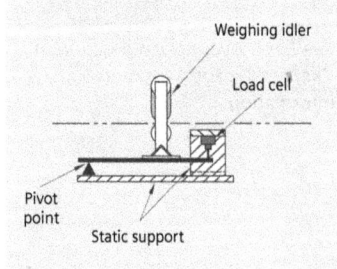

See *Basic Pivoted Design* for more information.

Performance rating: GOOD

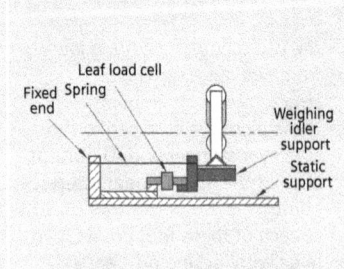

See *Leaf Spring as a Lever* for more information.

Performance rating: GOOD

SIBS[†]: load directly applied to a single dual-beam single-point load cell	SIBS: load directly applied to two of dual-ended bending beam load cells
• load applied in compression at right angles to the direction of belt travel • fast reaction time • load cell is designed for platform use • stability a concern with only one centrally located load cell – large platforms with only one load cell force side rotation pressure on load cell	• load applied in compression to each end of the load cell, while the load cell is rigidly supported in the middle • fast reaction time • load cell used is not designed for platform use • larger idler sets may not be stable

See *Direct Application to Dual-Load Cell Design* for more information.

Performance rating: GOOD

See *Direct Application to Dual-Load Cell Design* for more information.

Performance rating: GOOD

[†]Single-idler belt scale

SIBS: load directly applied to two of the dual-beam parallelogram load cells	SIBS: load directly applied to two of its triple-beam parallelogram load cells
• load cells designed for platform use • very fast reaction time • good stability • no moving parts • no hysteresis • low deflection	• load cells designed for platform use • very fast reaction time • good stability • no moving parts • no hysteresis • very low deflection • ignores interference from other stress forces

51

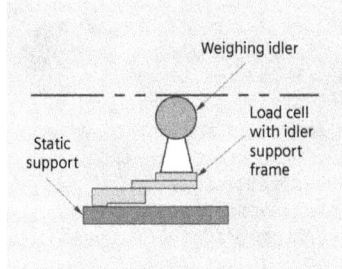

See *Direct Application to Dual-Load Cell Design* for more information.

Performance rating: VERY GOOD

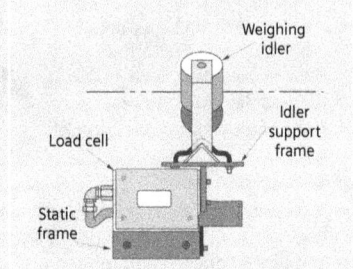

See *Direct Application to Dual-Load Cell Design* for more information.

Performance rating: EXCELLENT

Common full construction single-idler belt scale designs

Type	Pivoted: load cell in tension or compression	Flexure bearing with idler mounted at load cell	Direct application of two triple-beam compression load cells
Reaction time	slow	medium	fast
Deflection of dynamic frame	limited	generally limited	minimal
Structural stability	good	Inconsistent	excellent
Isolation of horizontal forces	good	generally good	excellent
Maintenance	bearings a concern	leaf spring can corrode	housekeeping only
Material buildup on dynamic frame	lots of surface for buildup, translates to zero shifting	lots of surface for buildup, translates to zero shifting	limited surfaces for buildup
Hysteresis	highest	medium	lowest
Ease of installation	difficult	very difficult, need added cross conveyor support	simple
Ruggedness	good	good	very good

Single-idler: full construction versus modular

Traditional belt scales were generally fully constructed; comprising a static frame and a dynamic frame which runs the width of the conveyor sensing the weight. However, modular belt scales have become increasingly more popular because of lower purchase costs.

Both of these belt scales use platform style, single-point load cells but the difference in the design lies in the support structure. The modular design has no support structure and the weight is transferred directly to the load cells, rather than onto a dynamic beam resting on load cells.

Full construction	Modular
Static structure Dynamic structure Load cells	Left weigh module Right weigh module
Applications	
• heavier duty applications • high accuracy requirement	• lighter duty applications • moderate accuracy requirement
Benefits	
• complete weighbridge is factory assembled • compensates for idler frames that have slight misalignment • simpler idler alignment • strengthens the conveyor • more accurate	• lower cost • flexible to different belt widths
Drawbacks	
• more expensive • made to one belt width each time • larger and more cumbersome	• only modules are factory assembled; cross conveyor construction is completed by the weighing idler • need to complete assembly in field without twisting between the two weigh modules • idler alignment more difficult • less accurate • conveyor structure is more critical because the individual units add no strength and integrity to the installation

53

Installing a modular scale

Installing a modular scale requires special attention to a number of details not required of a factory assembled, full construct scale made for a specific belt width. The assembly nature of the modular scale requires that special attention is paid to the following to ensure the greatest possible accuracy:

1. Location is strong enough to withstand rotational twist.
2. Keep vertical and parallel.
3. The conveyor stringer must be strong enough to withstand twisting.
4. Top of idler support bracket.
5. Keep idler support brackets on same plane and parallel to each other.
6. Bottom of module mounting flange.
7. Add cross structure, if necessary, strengthen conveyor and fix conveyor stringers.
8. Ensure proper alignment with adjacent idler rollers.

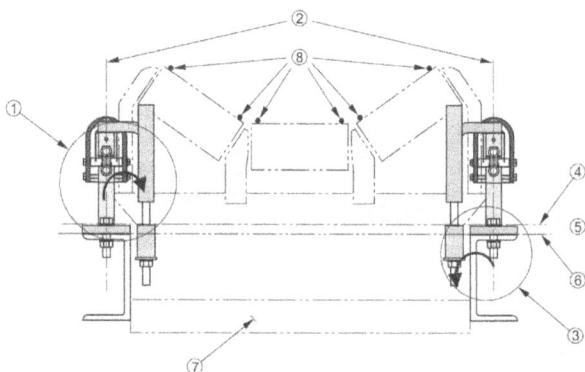

NOTES:
- Ensure that rotation side forces are not applied to the load cells because that will affect performance.
- Ensure there is less than one mV difference between the load cell output readings when installation is complete (belt lifted).

Common multi-idler weighbridge designs

Full construction multi-idler belt scales are still used despite the increasing popularity of single-idler weighbridges used in tandem. Based on the premise that the weighing length must be long for best accuracy, multi-idler weighbridges were the standard installation when accuracy was required.

However, multi-idlers require even more idlers to match the accuracy currently provided by the modular direct-to-load cell designs. Pivoted style multi-idlers are still useful for some extreme conditions such as very light loading, or very thick conveyor belting, as long as the multi-idler weighbridge selected has tare weight counter-balance features.

Pivoted two-idler and four-idler weighbridges with compression load cells

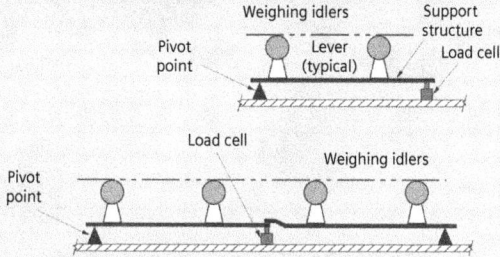

- one or two load cells depending on belt width
- slow reaction time as the lever rotates through the pivots
- reasonable stability due to application of load to load cell
- pivots, with wear and temperature variation, may also contribute to mechanical hysteresis

Performance rating: POOR

Pivoted two-idler and four-idler weighbridges with tension load cells

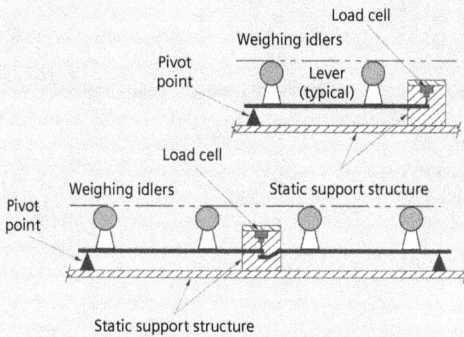

- one or two load cells depending on belt width
- slow reaction time as the lever rotates through the pivots
- good stability
- pivots, with wear and temperature variation, may also contribute to mechanical hysteresis

Performance rating: GOOD

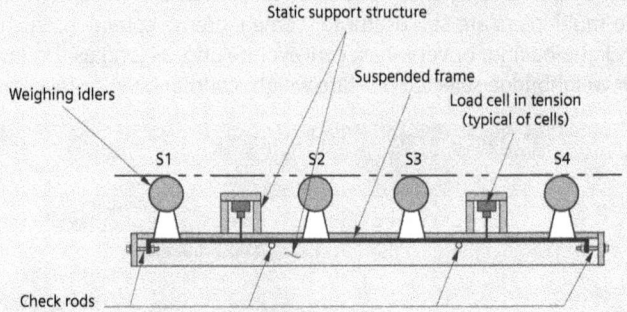

Fully suspended weighbridge with check rod stabilizers

- fast reaction time
- reaction time reduced by check rods† required for good stability

Performance rating: VERY GOOD

† Check rods are long bolts that check or eliminate horizontal movement but are sized to allow limited vertical movement.

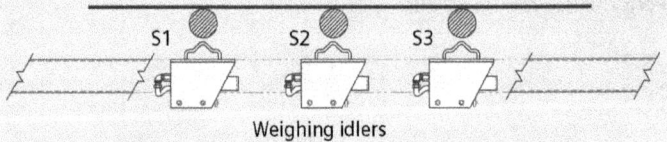

**An alternative to the above designs:
Two or more MSI single-idler belt scales in tandem**

- load directly applied to triple beam parallelogram load cells
- very fast reaction time
- good stability
- no moving parts
- no hysteresis
- low deflection

Performance rating: EXCELLENT

Full construction multi-idler designs versus multiple single-idler designs

Type	Pivoted: load cell in tension or compression	Fully suspended weighbridge with check rod stabilizers	Two or more MSI single-idler belt scales in tandem
Reaction time	slow	medium	fast
Deflection of dynamic frame	of concern	generally limited	minimal
Structural stability	fair to good	good due to check rods	excellent
Isolation of horizontal forces	good	generally good	excellent
Maintenance	bearings a concern	check rods may corrode	housekeeping only
Material buildup on dynamic frame	lots of surface for buildup, translates to zero shifting	lots of surface for buildup, translates to zero shifting	limited surfaces for buildup
Hysteresis	highest	medium, because of check rods	lowest
Ease of installation	difficult, due to size of weighbridge components	very difficult, weighbridge is very bulky and heavy	simple, limited manpower required
Ruggedness	generally good	good	very good

Multi-idler: full-length versus multiple single-idler in tandem

The multi-idler scale is used in high accuracy applications and where the conveyor belt has minimal flexibility. There are two approaches:

- the full-length multi-idler – it has a number of idlers on a weighbridge which then transfers the weight to the load cells. Thus while the multiple idlers provide a number of contact points for weight transfer, the data is still collected from a single set of load cells.

- multiple single idlers in tandem – collects the weight from each set of single scales, weighing the load more than once. The integrator software calculates the results from the multiple single-idler belt scales, effectively weighing the material being conveyed more than once and then providing the total.

Multiple single-idler belt scales installed in tandem provide the greatest accuracy and is a popular choice for numerous suppliers. Systems with two or three single idlers in tandem are field proven in NTEP[10], OIML/MID[11] and Measurement Canada[12] certified applications. The two/three MSI installed in tandem have even obtained +/- 0.125 % accuracy.

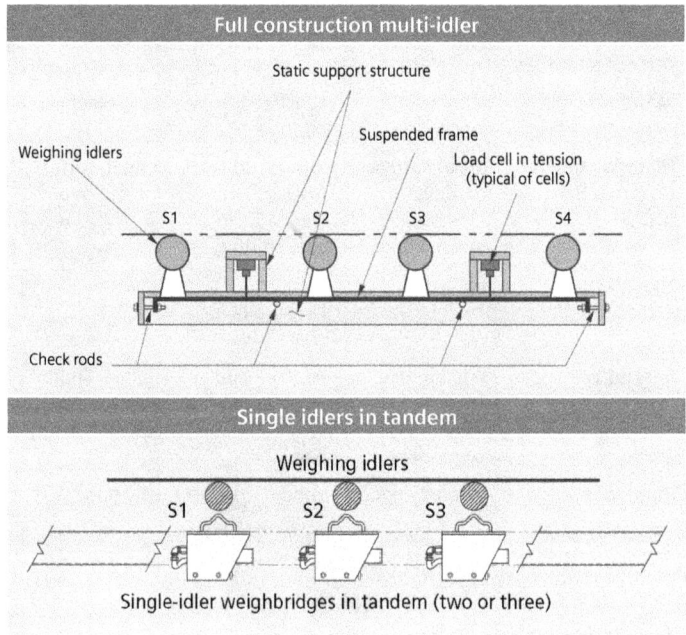

Full construction multi-idler

Single idlers in tandem

Single-idler weighbridges in tandem (two or three)

---

[10] NTEP is a standard for weights and measures used in the USA.
[11] OIML/MID are weights and measures standards used by the European Community.
[12] Measurement Canada is the government agency that controls weights and measures standards in Canada.

Full construction	Single-idler, tandem
• Heavier duty applications • High accuracy requirements	• Heavier duty applications • High accuracy requirements
Benefits	
• Factory assembled • Less assembly in the field	• Lower cost • More modular installation
Drawbacks	
• More expensive • Larger, heavier, and more cumbersome • More difficult to install due to size and weight, require lifting machinery • Not true "free floating" weigh idlers because they require check rods for suspended frame stability	• None noted

The MSI – advanced weighbridge design

In the 1980s, the MSI[13], with a new weighbridge design, was intro-duced by Siemens Milltronics, bringing an accurate and affordable single-idler belt scale to the market. This new design originally used two dual-beam parallelogram platform load cells, one at each extreme end of the belt scale, and balanced them electronically; thereby creating one large weighing platform. With this load cell design, the weighing idler is applied directly to the load cells with no pivots or levers. Because the load cells are located so near the conveyor stringers, the structural integrity of the belt scale is only critical in the area of each load cell, providing fast reaction to load changes resulting in greater accuracy.

The performance advantage is based on the reaction time of the weighbridge. Because it directly influences the load cells, the MSI design requires half of the retention time than the traditional pivot-ed belt scale, leading to a quick and accurate response.

---

[13] Milltronics Single Idler

To demonstrate the high performance output of the MSI we took a pivot point system and dropped a weight from a fixed height in two spots. The first (1) was between the pivot and the load cell and the second (2) over the load cell. Then we dropped a weight on an MSI system which measures direct load cell contact (3). All three impacts on the load cell were measured using an oscilloscope, with (r) as the response time. As you can see, the MSI design had the quickest reaction time, ensuring the greatest accuracy.

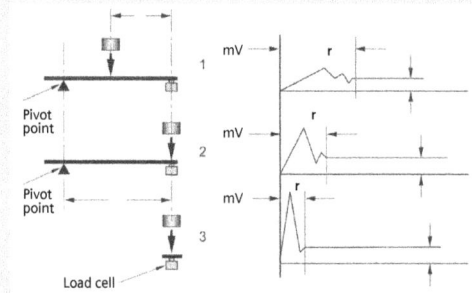

This unique design is available exclusively from Siemens and has a global reputation for accuracy, quality, and durability. While other manufacturers have adopted the two load cell concept, their designs do not match the accuracy performance of the MSI because their load cells are not truly suited to the parallelogram design where the weighing idler is directly applied to the load cells.

Approvals

Hazardous:

- suitable for CSA / FM, IEC Ex and ATEX hazardous approvals

Trade:

- MSI: approved and certified in the field by Measurement Canada, MID (to OIML standards) in Europe, SABS in South Africa, CMC in China, and GOST in Russia
- MMI-2 (two MSI in tandem) has been type approved and certified in the field by Measurement Canada, MID (to OIML standards) in Europe, NTEP in the USA, CMC in China, and GOST in Russia

> **NOTE:** Since this design was first introduced, the dual-beam parallelo-gram load cells have been replaced by state of the art triple-beam stain-less steel load cells with redundant protection from moisture ingression and closer sensitivity tolerances. For more on load cells, please see *Chapter 5.*

The design

Using single-point loads cells[14] allows for the implementation of a weighbridge that reads accurately across the width of the conveyor belt (platform effect), especially if electronic load cell balancing is used.

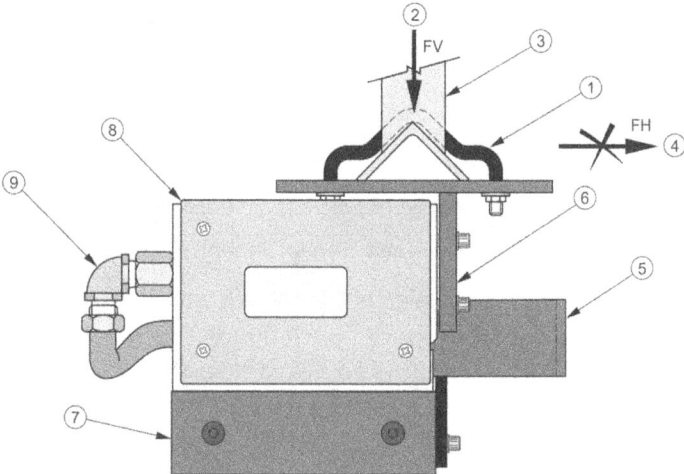

1. Idler clamp
2. Measures only vertical force
3. Belt conveyer idler frame
4. Horizontal force ignored
5. Calibration weight bracket
6. Heavy steel idler frame support
7. Structural steel, C channel
8. Stainless steel triple-beam load cells (2)
9. Load cell cable conduit

> **NOTE:** Hanger mounting bracket not shown.

With a triple-beam parallelogram, single-point load cell only the vertical forces associated with conveyor loading are measured. The horizontal forces are nullified by the horizontal structural beams of the load cell.

---

[14] Also known as platform load cells.

Vertical force

overload

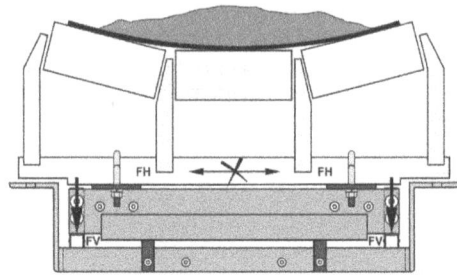

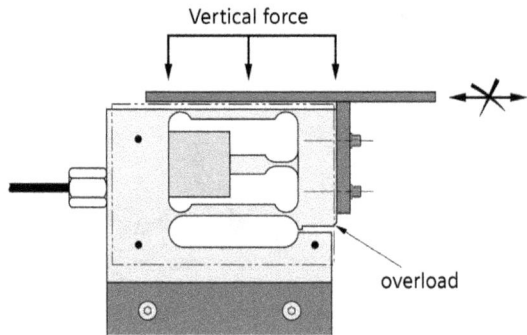

Legend	
t/h	Metric tons per hour
STPH	short tons (Imperial) per hour
m/sec	meters per hour
fpm	feet per minute

Upgrades to the initial design of the MSI belt scale have increased its performance capacity:

- upgraded load cell capacity (to include up to 1500 pound load cells)
- design rate (increasing capacity from 5,000 t/h (5,500 STPH) to 12,000 t/h or (13,225 STPH)
- high belt speeds can easily be tolerated (up to 5.0 m/sec or 1000 fpm) with some belt speeds coming in above 6 m/s (1200 fpm)

I was at a customer site and asked the operator how he liked the low cost scale he was using. "I like it so much that I buy another one every year," he replied. I was impressed and asked how many scales he had in operation. It turns out, there was only the one and he replaced it with the same model every year.

I showed him Siemens Milltronics scale applications that were still in service after 30 years, and he bought an MSI on the spot.

Designed for exceptional accuracy

MSI single-idler belt scale

The single-idler MSI system is as accurate as, or better than, any single-idler or any two-idler pivoted design belt scale.

- load directly applied to triple beam parallelogram load cells
- very fast reaction time
- good stability

Accuracy: +/- 0.5%, or better, of totalized weight over the range of 20% to 100% of design rate.

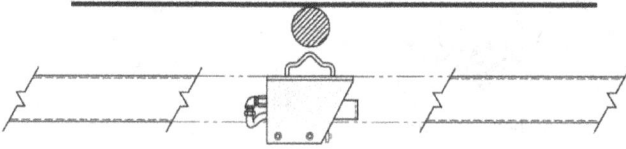

Double-idler MSI belt scale

A two-idler MSI system is as accurate or better than any four-idler pivoted or fully floating belt scale design.

- two MSIs applied in tandem creates the MMI-2 multi-idler belt scale
- very fast reaction time
- good stability

Accuracy: +/- 0.25%, or better, of totalized weight over the range of 20% to 100% of design rate.

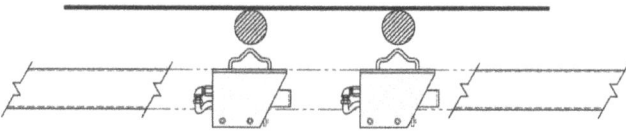

Triple-idler MSI belt scale

A three-idler MSI system is as accurate or better than any belt scale available.

- three MSIs applied in tandem to create the MMI-3 multi-idler belt scale
- very fast reaction time
- good stability

Accuracy: +/- 0.125%, or better, of totalized weight over the range of 25% to 100% of design rate.

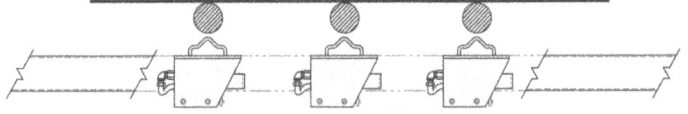

MSI on belt conveyer

MMI-2 on belt conveyor

Summary

Belt scale designs are numerous and are available in both single and multiple idler construction. However, the MSI with its direct load cell contact and rugged assembly provides the most accurate and repeatable measurement. Furthermore, its modular design provides even higher accuracy when it is used in multiples, effectively replacing rigid multi-idler belt scales.

Chapter 5

Load Cells

This is ourselves / Under pressure.[1]

While the design of the belt scale is crucial to its performance, it is the load cell that turns the sensed weight into the electrical signal used in the measurement.[2] This device comes in several iterations:

- **Linear Variable Differential Transformers (LVDT):** these load cells measure a displacement proportional to the forces applied to the belt scale weighbridge. Producing a variable low voltage AC signal proportional to movement, they operate with a fixed excitation voltage and a fixed frequency.
- **Strain gauge:** these load cells are the most commonly used in either compression or tension mode, and can have either digital or analog output.
- **Vibrating wire:** these load cells use a load sensing vibrating wire which changes frequency when tension forces are applied. Used in small weigh feeders by some manufacturers, they are rarely used with belts scales because of high cost considerations.

NOTES:
- load cells based on pneumatics and hydraulics, although used only to a minor degree on older weighbridge designs for hazardous rated areas, are not covered here
- capacitance load cells are now available, but because they are not yet used with belt scale weighbridges, they are also not covered

---

This chapter discusses linear variable differential transformers (LVDT) and strain gauge load cells used in either tension or compression. Tension and compression forces are best illustrated by a coil spring. Compression forces occur when that spring is squeezed shut, and tension forces occur when the spring is pulled further apart. The strain gauge works exactly the same way as the force of the weight compresses the unit or the release of weight creates tension forces.

---

[1] Queen, "Under Pressure." *Hot Space,* 1982.
[2] Load cells can be used independent of scales also, for example, to calculate tank level by weight.

NOTE: LVDT and strain gauge load cell designs are presented here in their most common forms as they are the popular technology. Other technologies are not discussed.

Topics

- Load cell technologies
- Strain gauge load cell types

Load cell technology

There are two main load cell technologies:

- LVDT
- Strain gauge

LVDT load cells

Linear variable differential transformers were popular in early electro-mechanical belt scales because the design is very robust and mechanical overload is almost impossible.

LVDTs provide a frictionless mechanical to electrical transduction; they read deflection from conveyor loading through the movement of a transformer core inside the bore of a cylindrical transformer. The primary windings are excited with a fixed frequency and voltage and the output of the secondary coils, proportional to movement of the core (a.k.a. *armature*), is measured in AC volts.

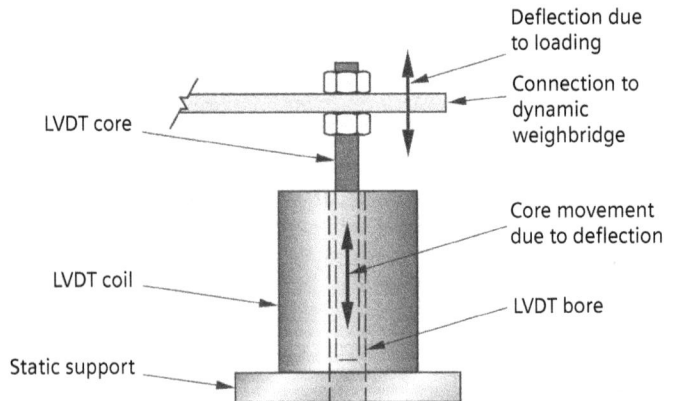

LVDT linearity curve

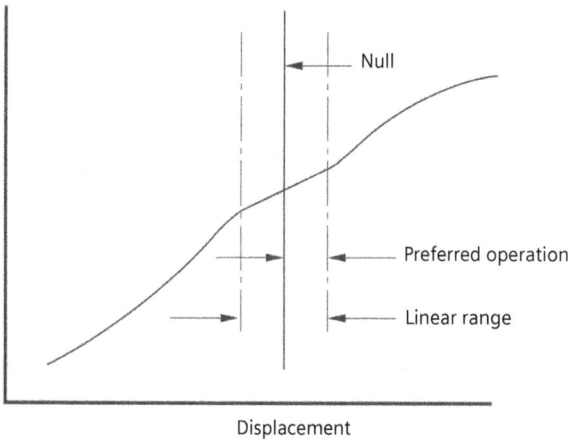

Null

Preferred operation

Linear range

Displacement

However, this design requires a large amount of deflection so it can present a good usable signal, unlike strain gauge load cells. A belt scale using an LVDT usually has problems with excess weighbridge deflection greater than 0.4 millimeters (1/64") because the LVDT may be using up to 2.0 millimeters displacement for preferred operational displacement as shown above in the linearity sketch.

> **NOTE:** Excess weigh idler movement greatly reduces accuracy. To achieve suitable movement, without excess deflection, the weighbridge design requires pivots and levers, and coil springs to stabilize the movement and limit weighing idler travel. All these features slow the reaction time. These problems with LVDT deflection have led most belt scale suppliers to use strain gauge load cells instead.

Strain gauge load cells

The basic sensor on a strain gauge load cell is a copper foil strain gauge. When tension forces (stretching) are applied to such an electrical conductor, its cross-sectional area decreases, causing its resistance to increase. When compression forces are applied to the conductor, the cross-sectional area increases (compressing) and conductor resistance decreases.

Developed some 60 years ago, the foil strain gauge has been refined over the years to suit the needs of modern force and load sensing. The strain gauges are placed at selected locations on the load cell element (body), creating a resistive circuit much like a

Wheatstone Bridge.[3] The circuit is adjusted to provide a 0 mV signal output at no load applied with a fixed excitation voltage supplied to the bridge. The load forces can be applied in tension, or in compression, in order to distort the balance of the bridge which provides an analog mV DC output proportional to the force applied.

Through computer generated finite element analysis, which looks at the total stresses and strains resulting from forces applied to the load cell structure, the modern load cell designer can determine the best location for the strain gauges on the machined load cell structure.

Bonded strain gauge

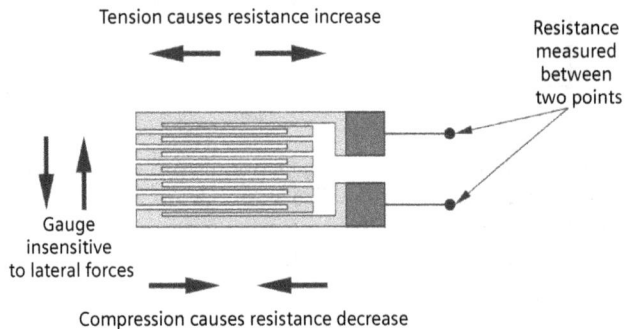

Tension causes resistance increase

Resistance measured between two points

Gauge insensitive to lateral forces

Compression causes resistance decrease

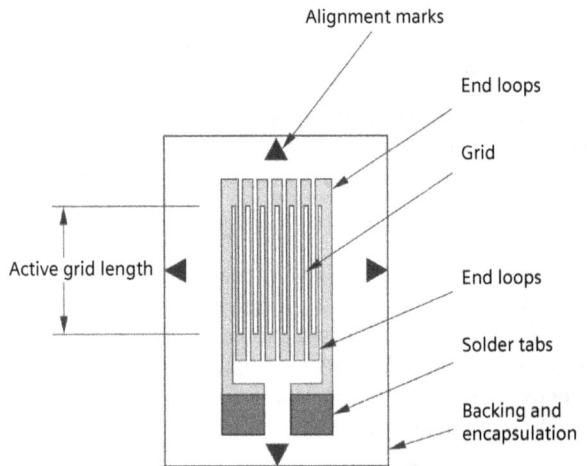

Alignment marks

End loops

Grid

Active grid length

End loops

Solder tabs

Backing and encapsulation

[3] A Wheatstone Bridge electrical circuit measures an unknown electrical resistance through balancing two legs of a bridge circuit, one leg of the circuit includes the unknown resistance.

Here is how they work:

- the strain gauges are glued in place at selected locations on the load cell element and interconnected
- this creates a resistive circuit similar to a Wheatstone Bridge
- the circuit is adjusted to provide a 0 mV signal o/p at no load. Force, or load, can be applied in a positive or a negative manner to provide deviation from zero as either tension or compression forces are applied to the strain gauges.
- the load cell is calibrated to a set sensitivity in terms of mV/V[4]

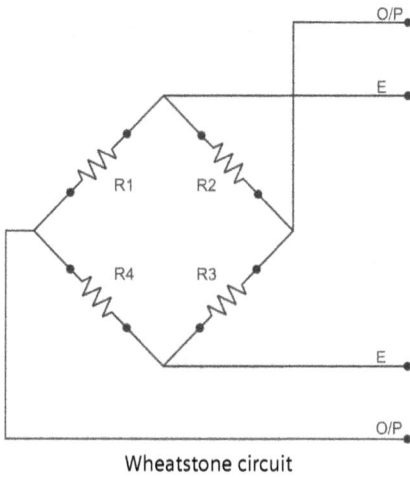

Wheatstone circuit

Although other components may be inserted into the circuit to account for temperature compensation, zero adjustment, and cornering (creating a platform effect); the basic circuit is as a bridge consisting of four resistors.

Example:
- 100 kilo load cell
- excited with a stable 10 VDC power supply
- sensitivity is rated at 2.00 mV/V

Output calculation for full scale capacity:

2.00 mV/V x 10 VDC + 20 mV

- its output at 0 kilo applied would be a nominal 0 mV

Legend	
E	V DC excitation source
O/P	load cell output in mV DC

[4] Millivolts generated per volt of excitation for a full capacity reading.

Simple load cell: a simple cantilever design with two strain gauges on top which will be in tension and two on the bottom which will be in compression when the force (load) is applied in compression:

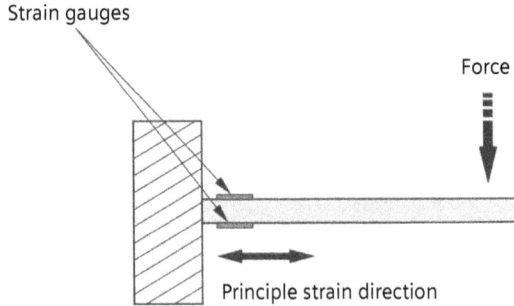

Strain gauges

Force

Principle strain direction

Basic strain gauge circuit with digital output

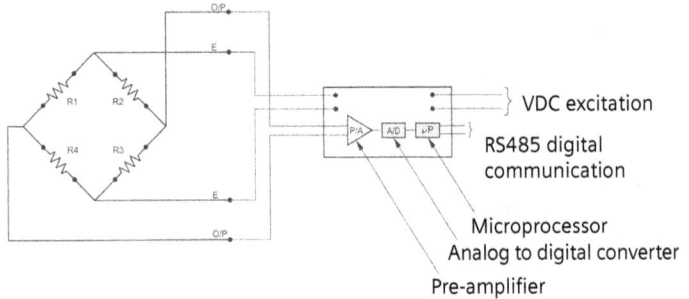

VDC excitation

RS485 digital communication

Microprocessor
Analog to digital converter
Pre-amplifier

EMI = Electromagnetic interference

While an analog mV signal has been the traditional load cell output, the digital world has improved its performance. An analog to digital (A/D) converter mounted internally to the load cell provides digital output, reducing the need for A/D circuitry in the interfacing belt scale integrator or other devices. Digital output provides higher resolution and greater immunity from EMI.

Industrial communication requirements also kicked digitization up a notch with the addition of a microprocessor and communication protocols like ModBus RTU.[5] Now the load cells talk to the belt scale integrator, or other devices, via bus communications.

Protecting the load cells

Load cells often operate under severe conditions and require a robust design to withstand the pounding. The following areas have significant impact:

- the adhesives bonding the gauges to the element surfaces
- tighter sealants
- improved potting compounds that protect the strain gauge circuit from moisture ingression
- encapsulating the strain gauges for protection from moisture and dust particle contamination

There are several common causes of load cell failure:

- moisture ingression
- over stress of the metal to which the strain gauges are bonded
- electrical overload – common causes are lightning strikes or welding[6]

A customer had to ship a belt scale system to a different site and Siemens Milltronics assisted with the shipping preparation because the conveyer was hinged and the belt scale would be upside down in transit. The belt scale was to be locked down to prevent load cell damages in transit. However, the operator forgot to lock it prior to packing up and moving the conveyor.

Traveling upside down on rough roads without any restraint caused the belt scales load cells to break apart. At the new site, the operator realized his error and attempted to weld the load cells back together.

While the load cells looked whole, they were far from operational. The strain gauges and various interconnection wiring were burnt off the load cell surfaces by the welding.

---

[5] While several different standards are used for bus communications, the flexible Modbus RTU is generally the preferred one.
[6] Be sure to design installation with overload protection which includes protection within the plant from lightning generated surges.

Load cell types

Low profile canister

This design (aka: pancake load cell) is one of the first used with belt scales, either with tension and in compression. General usage, however, is the compression mode with the weighbridge applied to the top side.

Typically, three designs can be applied to this load cell:

- a multi-spoke shear section
- multi-spoke bending section
- diaphragm bending section

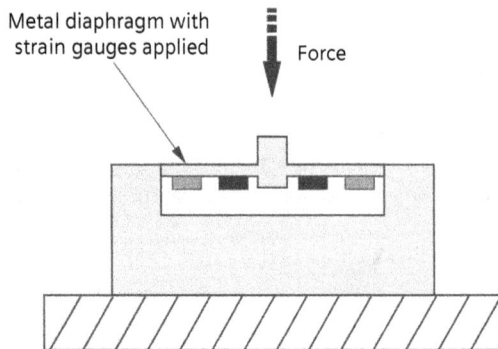

Each design requires near perfect load introduction so the load must be introduced through a flexible bearing point; this is only possible if the weighbridge is stabilized mechanically with pivots/ levers and/or check rods.

Drawbacks to this design and the use of mechanical parts:

- introduction of hysteresis to the weighbridge operation
- increased reaction time to account for the transfer of the vertical forces to the strain gauges, slowing the weighing process
- friction on the load button surface from variable surface movement can contribute to non-repeatable readings

High profile canister

This load cell (aka: column load cell or multi-column load cell) is one of the first used with belt scale designs. It is used in compression mode with the weighbridge applied to the top. Some models only have a load button for compression mode operation.

Strain guage locations

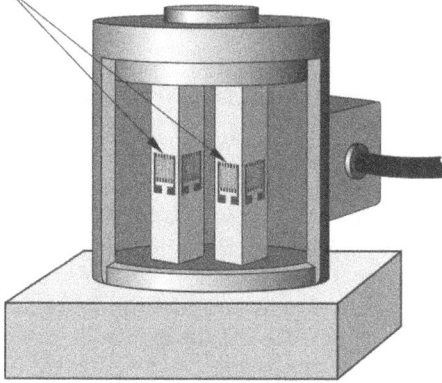

This load cell requires a purely vertical load, right on the center of the button. It needs a flexible bearing point with the weighbridge stabilized with pivots/levers and/or check rods, the long bolts that check horizontal movement of the weighbridge but flex to allow vertical movement.

Single-ended double-bending beam

This load cell is used for low capacities and performs with good linearity (linear relationship between forces applied and signal output). The strain gauges are bonded on the flat upper and lower sections of the load cell at maximum strain points. One disadvantage is that it must be loaded correctly to obtain consistent results.

A bending beam (aka: cantilever beam) can be used in tension by the weighbridge structure pulling down, or used in compression with the weighbridge applied to the top side. Under tension, the load is applied through a flexible wire rope. Compression requires a flexible bearing point.

Strain gauges

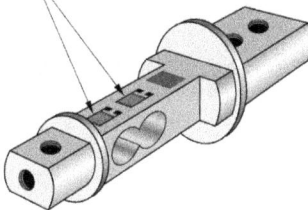

Pivot points/levers and/or check rods are required in this design and have the following negative impact:

- introduction of hysteresis to the weighbridge operation
- increased reaction time to account for the transfer of the vertical forces to the strain gauges, slowing the weighing process

S beam in tension

The S beam can be used in either tension or compression mode, but for belt scales, the tension mode is most effective. Hanging the weighbridge floating frame from the bottom of the load cells provides faster reaction to loading without load cell compression button loading errors. The S cell is also less affected by off-center loading than the canister type.

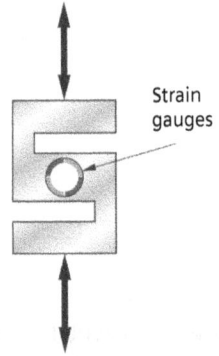

Strain gauges

However, the necessary flexible connection, often a wire rope, requires the weighbridge design to include stabilizing components like check rods or pivot points.

This design requires pivot points/levers and/or check rods, which have the following negative impact:

- introduction of hysteresis to the weighbridge operation
- increased reaction time to account for the transfer of the vertical forces to the strain gauges, slowing the weighing process

Double-ended double-bending beam

This load cell is used for low capacities and performs with good linearity. It must be loaded correctly to obtain consistent results. With this design, the strain gauges are bonded on the flat upper and lower sections of the load cell, at two different locations, at points of maximum strain.

A double-ended bending beam is generally used in compression with the load applied to the two ends. The static support is applied at the mid point. The load cell can support a frame supporting a single conveyor idler.

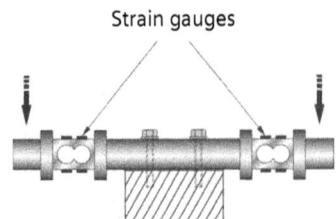

Strain gauges

74

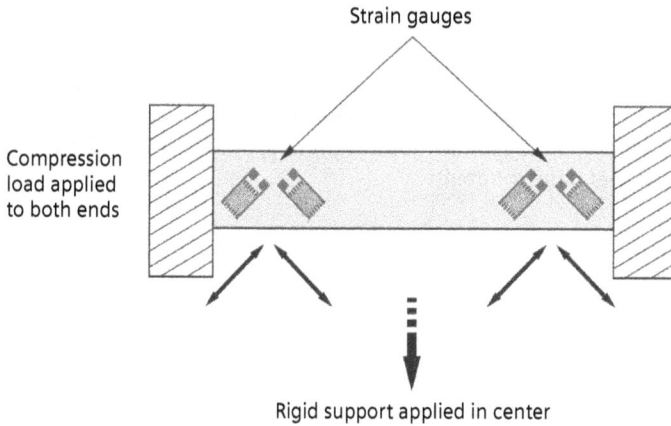

Strain gauges

Compression
load applied
to both ends

Rigid support applied in center

Dual-ended shear beam

The dual-ended shear beam is supported on one end with the load applied to the opposite end of the load cell. These load cells come with self-checking mounting hardware and can be rigidly bolted to the static and dynamic structures without requiring the stabilization provided by pivots/levers and/or check rods. However, in extreme applications such as large pan conveyor applications, check rods can ensure structural stability.

Because the load is directly applied to the load cells, the strain gauges respond quickly to loading changes, improving accuracy. Horizontal forces are eliminated from the weighing process by the structural beams of the load cells. Siemens uses this load cell in a heavy-duty four-load cell scale installed on pan conveyors and apron feeders.[7]

Flexibility in the load cell mounting hardware allows for thermal elongation of the weighbridge structure without it affecting weighing accuracy. These four load cells can also support a long weighbridge which itself supports as many as four idlers.

These load cells are available in nickel-plated steel and in stainless steel, and the strain gauge seal can be had with a welded stainless steel ring. This encapsulated design is suitable for wash down applications.

---

[7] For more information, see *Chapter Two*.

Dual-beam single-point

These single-point platform load cells were originally designed for small static scales such as postal scales and grocery scales. Used in the compression mode, they are designed, built, and factory adjusted to provide repeatable readings over a specific platform size when loaded at different points.

Legend	
T	strain gauge in tension
C	strain gauge in compression
FC	compression force (or load)
FT	tension force (or load)

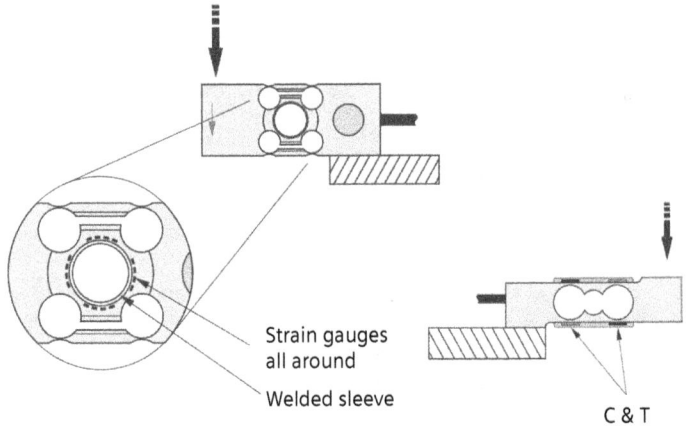

Strain gauges all around

Welded sleeve

C & T

Two of these load cells will directly support a conveyor idler or slider belt weigh pan. They can be rigidly bolted to the static and dynamic structures without needing the stabilization of pivots/levers and/or check rods.

Because the load is directly applied to the load cells, the strain gauges respond quickly to loading changes, greatly improving accuracy. Horizontal forces are eliminated from the weighing process by the structural beams of the load cell.

FC

The load cells are available in aluminum, nickel-plated steel, and stainless steel. Stainless steel versions can have the strain gauge area sealed with a welded stainless steel ring, creating an encapsulated design suitable for wash down applications.

Rigid support

Triple-beam compression single-point

The triple-beam compression strain gauge load cell is the state-of-the-art load cell for belt scale design. Its unique parallelogram arrangement allows it to integrate directly into the weighbridge structure.

Fast load sensing occurs through the direct application of the vertical load to the load cell while the structural beams of the load cell greatly reduce the horizontal forces.

The triple-beam design avoids a common measurement error attributed to dual-beam single-point load cells which are designed with top and bottom flexures. The strain gauges are placed on the top and bottom flexures where they measure the tension and compression strains that result from the deflection of the load cell. When a horizontal force is created by the movement of the weighbelt or because the load is applied off center, the horizontal force is super imposed onto the proper bending stress measured as a result of the load. Mixing these forces causes measurement error.

The triple-beam design avoids this mixing-of-forces error by measuring the stresses that result from vertical forces only on a third beam, located between the two outer beams (or flexures). The strain gauges in the center beam provides much better performance than the traditional dual-beam design.

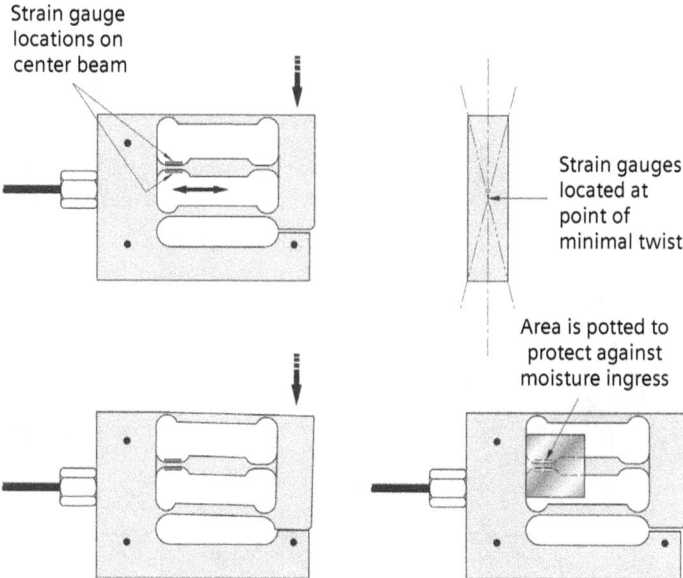

Strain gauge locations on center beam

Strain gauges located at point of minimal twist

Area is potted to protect against moisture ingress

Load cells at a glance

Comparison of load cells

Type	LVDT	Canister	S beam	STRAIN GAUGES			
				Single ended double bending beam	Dual ended shear beam	Dual-beam single-point	Triple-beam single-point
Pressure mode	Deflection	Compression	Tension	Compression	Compression	Compression	Compression
Moisture protection	Generally good (with cover)	Excellent	Generally good	Generally good	Generally good	Generally good	Excellent
Parallelogram movement	N/A	No	No	No	No	Yes	Yes
Adjust platform	N/A	No	No	No	No	Yes	Yes
Resolution	Poor	Good	Good	Good	Good	Good	Excellent
Deflection	High	Low	Low	Low	Low	Low	Very low
Overload stop	Only with weigh bridge design	Only with weigh bridge design	Only with weigh bridge design	Only with weigh bridge design	Only with weigh bridge design	Only with weigh bridge design	Built in
Rugged	Very good	Good	Good	Good	Good	Good	Excellent

Balancing load cells is like using two bathroom scales to find out how much you weigh. When you stand on one scale, you get the total weight. When you place a foot on each scale, you get two readings. Unless you have your feet and weight perfectly balanced, those readings will not add up to the correct weight provided by the single scale. The individual scales need to be equally balanced.

The same applies when independent scales are used together in a weighing system. Material running on a belt conveyor often rides closer to one side of the belt, depending on factors such as material feed point conditions and orientation; multiple feed points, flow rate, belt tracking, temperature, and material size and shape. Properly balanced load cells will weigh repeatable no matter what the side to side position of the material on the belt.

Summary

The load cells in a weighing system are the transfer point where the load is converted to data for the integrator to calculate into weight. Choosing the correct load cell for the application and system is thus crucial to dependable and accurate performance. Siemens carefully matches the correct load cell with application so that the belt scale operates with optimal performance.

Chapter 6

Speed Sensors

Let it roll, baby, roll / Let it roll, all night long.[1]

Speed sensors play an integral role in the calculation of weight, and it is crucial to use a robust and accurate device. A belt scale system requires two data inputs to calculate accurately the weight of material moving along the belt. The load cells provide the weight reading of the material loading on the belt, and the belt speed detector provides the linear speed reading. Together, the integrator calculates the dynamic weight.

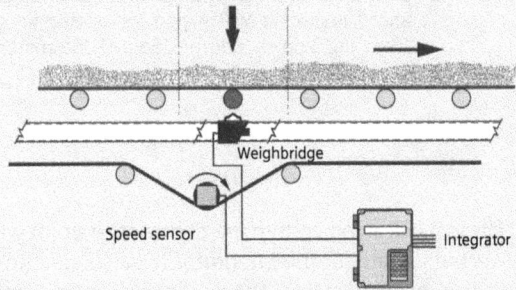

At times, belt speed detectors get lost in the presentation of a belt scale system because they are too often defined as only simple ancillary devices. However, they are critical to the success of a belt scale system and close attention must be given to select the most suitable type for the application.

Topics

- Belt speed detection
- Speed sensors

Belt speed detection

Speed sensors can take several forms and vary in installation location. Regardless of which speed sensor is used, the signal output is generally a pulse wave in one of two forms:

---

[1] The Doors, "Roadhouse Blues." *LA Woman*. 1971.

- a varying frequency and amplitude sine wave
- a DC pulse created with the reactive operation of an open collector transistor output

The most accurate speed measurement is derived from calculating the speed of a wheel or a pulley driven by the conveyor belt. There are two common methods used to read the conveyor belt speed:

- Belt speed sensor – rests on the belt and measures its speed directly
- Pulley driven sensor – directly attached to a shaft on one of the pulleys on the conveyer

NOTE: The constant speed function of the integrator is sometimes used for the weight calculation while other installations use the drive motor RPM as a belt speed indicator. However, while the drive motor operates at a constant speed, the belt speed may vary, even on constant speed belt conveyors. Belt speed is influenced by load applications and by belt stretch (due to loading and temperature). For greatest accuracy, the belt itself needs to be measured and timed in order to verify its actual speed.

Return belt speed sensor (RBSS)

Easily installed close to the belt scale assembly, the RBSS provides a signal generated as the wheel on the sensor rotates on the return belt. To secure this cost-effective unit in place, position a cross bar between stringers – either just before or after a return belt idler, or use a mounting bracket. The weight of the RBSS ensures positive rotation of the wheel in the middle of the return belt, and pulses from the magnetic sensor are generated by the rotation of the 60-tooth speed sprocket driven by the wheel.

The sensor is mounted between the conveying strand and the return strand of the belt so that the wheel is driven by the return belt travel.

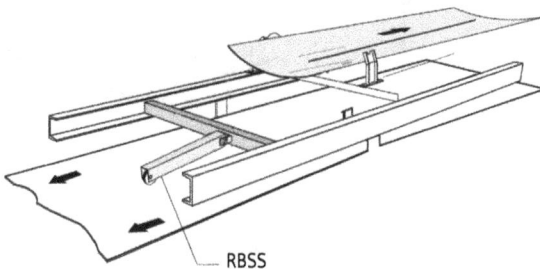

RBSS

Benefits:
- Lower purchase price
- Easy to install
- Easy to locate adjacent to the belt scale weighbridge

Drawbacks (model dependent):
- Low resolution per revolution
- Possibility for wheel slip on wet belt
- Possibility of wheel bounce if return belt bounces
- Short life with bearings and wheel, especially if not aligned to belt
- Jammed wheel may wear belt

A proximity switch generates pulses proportional to the wheel rotation, employing either inductance or magnetic sensors to react to defined targets on the wheel drum, or on a wheel-shaft mounted toothed sprocket. Less rugged units often use a rotary encoder driven by the wheel shaft to create a high resolution signal; however, the risk of damage to the encoder from moisture, dust, and corrosive material is higher.

Pulley driven belt speed sensors

A common form of conveyor belt speed detection uses an existing tail pulley,[2] a snub pulley, or a bend pulley driven by the return belt strand and applies a cantilevered-shaft mounted sensor to the pulley shaft.

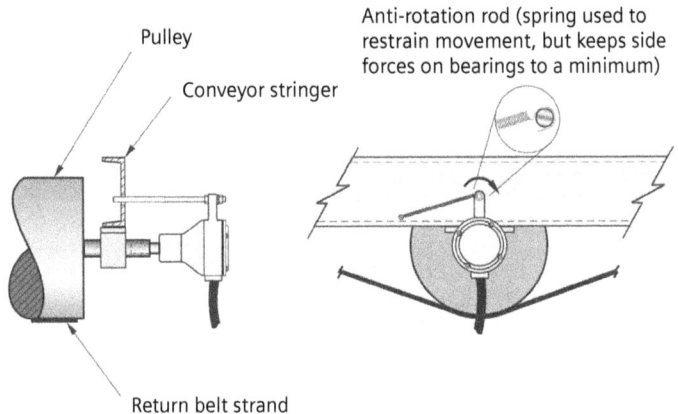

Pulley

Conveyor stringer

Anti-rotation rod (spring used to restrain movement, but keeps side forces on bearings to a minimum)

Return belt strand

---

[2] A tail pulley is at the tail of the belt conveyor opposite the normal discharge end. Snub pulleys are installed in conveyors to get a higher angle of wrap on the drive pulley. A bend pulley changes the direction of belt travel.

- Driven pulleys with a shaft-driven speed sensor generally offer better speed data because they are slip free
- When preparing a pulley shaft for mounting make sure the clearance hole for the speed sensor shaft is concentric with the shaft's outer diameter
- Calculating the value of the speed sensor factor is done either in pulses per foot or pulses per meter using the number of pulses per revolution divided by the circumference (diameter × Pi) per revolution of the driven pulley.

$$\text{Pulse/foot} \ = \ \frac{\text{pulses per revolution}}{\text{Dia. (in ft)} \times \text{Pi}}$$

$$\text{Pulse/meter} \ = \ \frac{\text{pulses per revolution}}{\text{Dia. (in m)} \times \text{Pi}}$$

NOTE: If a suitable pulley is not available, a bend pulley can be installed to mount the speed sensor. These bend pulleys generally come modified for the mounting of the speed sensor.

A cement plant had problems with a weighfeeder. It first worked fine and then inexplicably developed problems when drawing material from the silo above it – it would speed up and slow down as if hunting for the correct control speed to satisfy the rate control set point. The result was an erratic rate display.

A possible and costly solution to redesign and rework the chute was proposed until a Siemens technician put an oscilloscope on the speed input of the belt scale integrator connected to the weighfeeder. Noting how the speed signal varied, even when the motor speed control was manually set at a constant speed, he checked the speed sensor and found that its shaft did not always turn at the same speed as the tail pulley shaft. Maintenance staff had failed to fully tighten a set screw on the tail pulley and the speed sensor shaft would slip at times. This affected the rate calculations, causing the control system to hunt for the ideal speed to deliver the desired feed rate. A few turns of a set screw saved thousands of dollars!

Proximity switch sensors

The Siemens WS100 is a shaft-mounted sensor that uses ferrous targets detected by magnetic proximity switches, creating a pulse train proportional to the belt speed. The pulley shaft drives the sensor shaft, and the pulley shaft can be drilled to accept the sensor shaft.

Another installation couples the two shafts magnetically. Make sure the body of the sensor is tightly tethered so that severe shocks do not free the sensor from the pulley shaft. The magnetic coupling also allows self centering of the sensor shaft onto the pulley shaft.

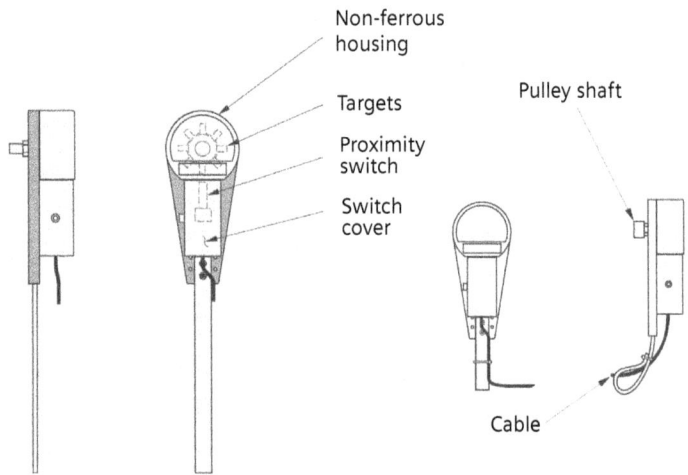

Non-ferrous housing

Targets

Proximity switch

Switch cover

Pulley shaft

Cable

If an Intrinsically Safe (IS) proximity switch and an IS switch isolator are used, the speed sensor can be used in hazardous applications.

Photocell toothed-wheel pickup sensor

This photocell sensor design became popular with belt scales systems when they shifted from mechanical to electronic. Initially the sensors used only one optical pickup; however, vibration could cause false speed signals at low rotational speeds. Siemens Milltronics, however, developed sensors with two optical pickups and added suitable logic to eliminate false pulsations caused by vibration. This rugged industrial design only allows for unidirectional sensing with a choice of either clockwise or counter-clockwise rotational output.

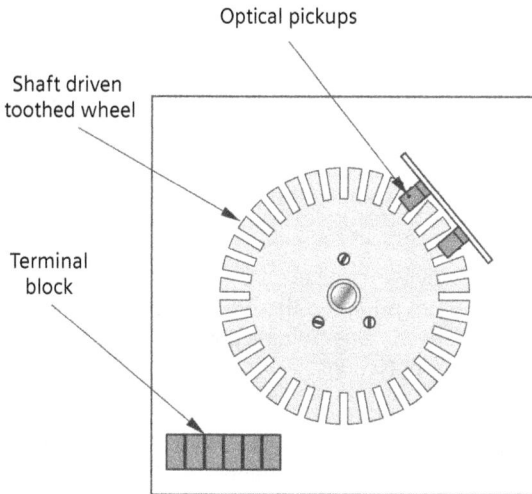

Optical pickups

Shaft driven
toothed wheel

Terminal
block

The resolution, however, is limited by the number teeth on the disc. A total of 36 is generally the maximum number of teeth possible on the wheel.

This design fits most belt conveyors and higher speed weighbelts. It's not suitable, however, for very slow moving conveyors or weigh-feeders where higher resolution encoders are required to achieve proper speed signal frequency.

Rotary shaft encoder sensors

These light duty sensors are generally used in secondary industries[3] where plant environments are cleaner and less rugged. They usually have quadrature outputs,[4] and interfacing logic circuits are required to reduce error potential caused by vibration. The higher the resolution of the encoder, the greater the potential for error from vibration interference from surrounding equipment. High resolution encoding thus needs suitable circuit interfacing logic installed between the encoder and the integrator.

[3] Primary industries refers to those harvesting raw materials (mining, cement). Secondary industries use refined materials in production (plastic pellets, pulp and paper, food).
[4] Two out of phase outputs. See *Appendix A: Glossary* for more information.

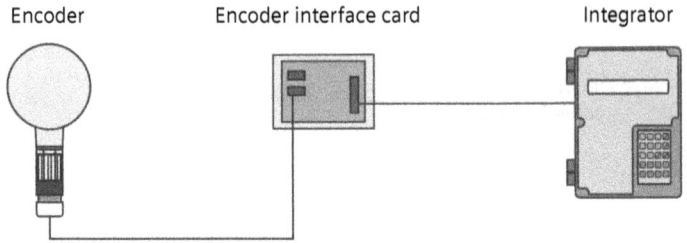

The integrator may already have this circuitry as part of its speed input, and it creates an anti-dithering logic so that only one direction is read as speed. This unidirectional reading keeps vibration from causing false pulsations.

This encoder interface card is used with a Siemens encoder on smaller weighfeeders like the Siemens WW100 and WW200, and is similar to the circuit used in the Siemens WS300.

High resolution encoder with anti-vibration interface circuit sensor

This advanced speed sensor design uses high resolution encoders with a robust mechanical protection and an electrical interface that eliminates false pulsation caused by vibration, even at encoder design resolutions of 2000 pulses per revolutions.

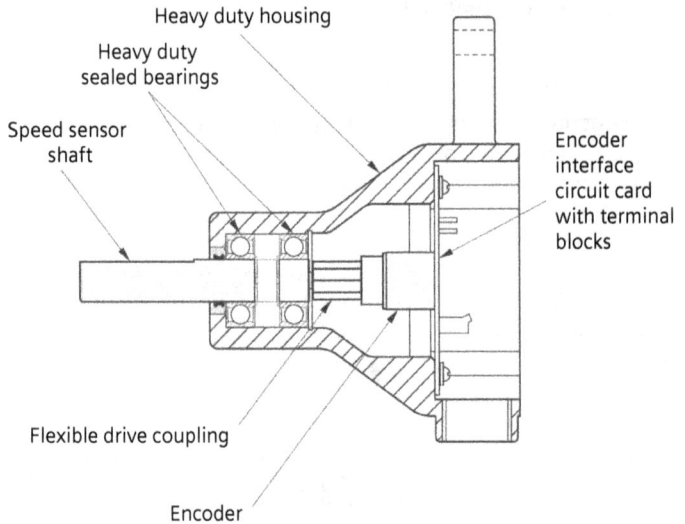

This design is similar to that of the rotary shaft logic circuit interface and typical of the SITRANS WS300 series of speed sensors, offering 32, 256, 1000, and 2000 pulse per revolution models.

SITRANS WS300 logic description

The WS300 rotary encoder produces quadrature outputs that are 90° out of phase from each other. These two quadrature signals (A and B) allow both the direction and the speed of rotation to be monitored.

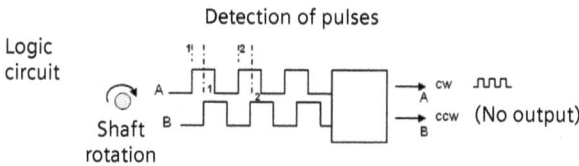

During the shaft's clockwise (CW) rotation, the A input leads the B input by 90°, causing the counter-clockwise (CCW) output to be set and held low (no output) while the A encoder input is clocked through to the CW output by B encoder input.

During counter-clockwise rotation, the opposite occurs. The B input from the encoder leads the A input by 90°, causing the CW output to be set and held low (no output) while the B encoder input is clocked through to the CCW output by the A encoder input.

The clocking of the encoder inputs greatly improves the reliability and quality of the encoder measurements by preventing false counts caused by vibration, noise, and jittering.

AC tacho-generator

Some belt scale systems use a shaft driven AC tacho-generator to provide a speed signal. No excitation voltage is required as the generator creates the AC output. With a diode across the input circuit of the integrator, the signal can be modified to a rough DC square wave. Both the frequency and the amplitude of the AC sine wave will increase with belt speed. However, the square wave may fall below the detection threshold at low rotational speeds leading to a speed signal loss.

Comparison of speed sensors

Type	Wheel driven w. proximity switch (Siemens RBSS and TASS)	Wheel driven w. shaft encoder	Pulley shaft driven encoder	Pulley shaft driven toothed wheel (Siemens WS100)	Pulley shaft driven techno generator	Pulley shaft driven encoder – separate vibration card	Pulley shaft driven encoder – anti vibration (Siemens WS300)
Resolution	Low to medium	Low to high	Low to high	Low to high	Medium	Low to high	Low to high
Reaction to vibration	Poor	Poor	Poor	Good	Good	Excellent	Excellent
Robustness	Good	Medium	Medium	Good	Good	Good	Good
Enclosure rating	NEMA 4X, IP65*	NEMA 4X, IP65*	NEMA 4X, IP65*	NEMA 4X, IP65	NEMA 4X, IP65	NEMA 4X, IP65*	NEMA 4X, IP65
Hazardous approval	Yes – with IS switch	Depends on encoder rating	Depends on encoder rating	Yes - with IS switch	N/A	Depends on encoder rating	CSA, FM, ATEX, IEC Ex Dust Approvals
Belt speed suitability	Medium to fast speed	Low to high	Low to high	Medium to high speed	Low to high speed	Low to high speed	Low to high speed
Risk of low speed signal error**	Low to medium	Very low	Very low	Low to medium	Low	Very low	Very low
NOTES:	* depends on ratings of proximity switch or shaft encoder ** caused by slippage, vibration, oscillation						

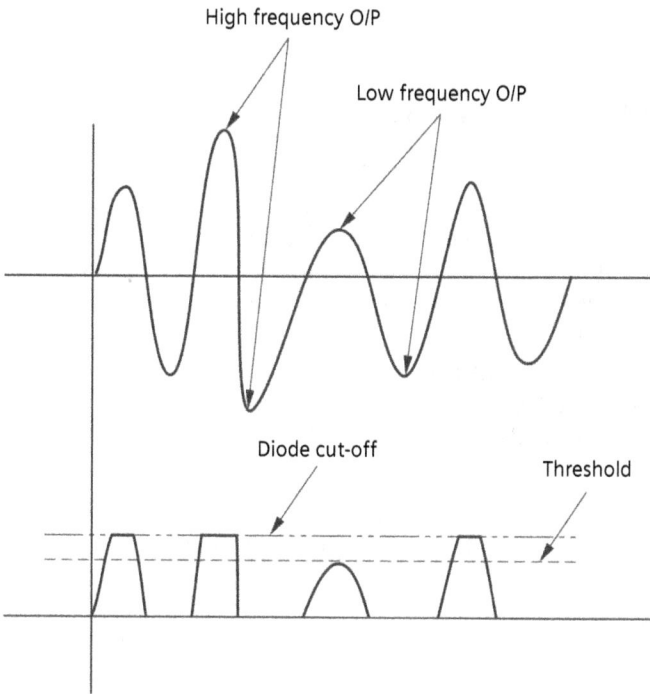

High frequency O/P

Low frequency O/P

Diode cut-off

Threshold

Summary

The speed sensor is an integral component to an effective belt scale system and it is important to choose the right one for the application. Whether it is attached to a pulley or riding the belt, its installation needs to take the application conditions into account to ensure the best performance.

Chapter 7

Integration

We've got to get together sooner or later . . . [1]

Belt scale integrators bring everything together into one happy data family by combining the weight signal from the weigh bridge proportional to belt loading and the belt speed to calculate the following:

- rate of material being conveyed
- the totalization of the same
- displayable values such as belt loading and belt speed

The primary calculation:

$$\boxed{\text{Rate} = \text{belt load} \times \text{belt speed}}$$

- belt load value is in kilograms per meter or pounds per foot
- belt speed is meters per second or feet per minute
- calculated rate is generally in kilograms per second or pounds per second. From this basic calculation, the rate is scaled to create readings in either tons, short tons per hour, or kilos or pounds per minute. The totalization can be in tons, short tons, kilograms, or pounds.

This calculation is done by a highly sophisticated electronic integrator which factors the mass of material and the speed it is moving.

Integrators like the Siemens Milltronics BW500 provide numerous additional features to enhance accuracy:

- support popular industrial communication buses to be a vital part of an industrial network
- patented load cell balance function eliminates matching load cells in belt scale assembly
- moisture and incline compensation

---

[1] Thunderclap Newman, "Something in The Air." *Hollywood Dream*. 1969.

- PID[2] function for rate control on shearing weighfeeders – where belt loading is relatively constant – but can also control pre-feeding devices
- may be used for ratio blending and controlling additives when operating in tandem with two or more weighfeeders
- batching, loadout, and alarm functions

In addition to the Siemens BW500, which is typical of stand-alone integrators, Siemens also offers the SIWAREX FTC PLC based weighing module designed to bring similar functionality integrated directly with a Siemens PLC system.

Topics

- Mechanical integrators
- Electronic integrators
- Integrator design
- User programming requirements
- Calibration
- Integrator packaging

Mechanical integration

In 1908, Merrick Scale developed the first belt scale system with a mechanical design that incorporated a rotating wheel. It tilted with the belt conveyor loading while being driven by a belt moving at a speed proportional to the conveyor belt speed. The change in wheel tilt and/or a change in belt speed caused a change in the wheel rotation speed and thus the rotation speed of the center shaft. This shaft's rotation was the integrated rate and it drove a mechanical totalizer, which recorded the aggregate of the material being conveyed.

A viscous fluid damper provided a dampening effect, as well as some tare weight counter balancing. This mechanical arrangement integrated belt load and belt speed, which is also the main function of current electronic belt scale integrators.

---

[2] Proportional Integral Derivative (PID) controller is a generic control loop feedback mechanism.

Tilt raise caused by beltload
gain increases rotational speed

Integrated rate
(turning of disc shaft)

Integrator wheel drive belt

Drive belt pulley (×4)

Integrator belt drive
speed proportional
to belt speed

Pivoted weighbridge

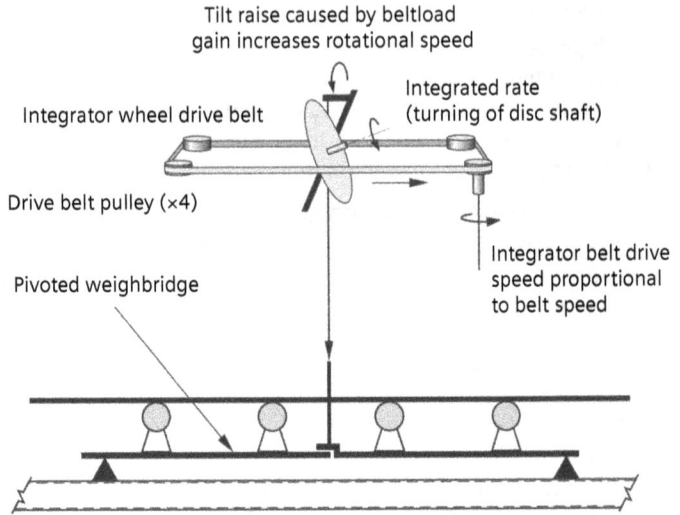

Another design option replaced the rotating mechanical totalizer with a small toothed wheel which was then sensed by a magnetic proximity switch, creating a pulse signal proportional to flow rate. The signal was then converted to provide an analog signal or to drive an electronic totalizer.

The belt scales with mechanical integrators were very popular and quite accurate, although they required a lot of maintenance. The natural wear of the mechanical parts required much servicing time, many replacement parts, and frequent re-calibration.

Electronic integrators

The first electronics based belt scale integrators were introduced in the 1960s, followed by microprocessor based designs in the 1980s. Subsequent decades introduced bus communications and digital data processing to make the current integrators into the sophisticated instruments operating today.[3]

---

[3] See *Chapter One* for more on the evolution of belt scale.

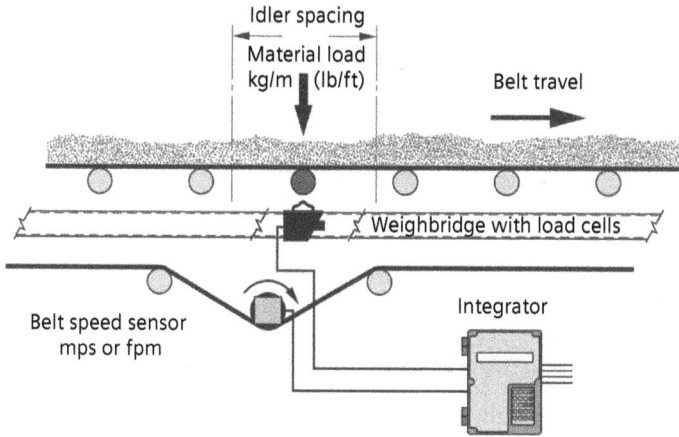

Basic function

The integrator receives load cell signals proportional to the conveyor belt loading and a speed signal proportional to conveyor belt speed.

The integrator does the calculation[4] and then formats it to display *rate, load, speed,* and *totalization*. The same information is also available in an analog signal form or a pulsed solid state relay form for use in remote totalization or sending data via an industrial communication bus.

Integrator design

Belt scale system integrators have turned the corner in the past few years from analog to digital devices. Although there is still some processing of the analog signals from the load cells required to prepare the signals for analog to digital conversion (A/D), as well as conversion from digital to analog (D/A) to create analog outputs, all calculations are performed digitally.

- Further development and the cost reduction of digital load cells will pave the way for further simplification of integrator designs and reduce the need for A/D conversion in the integrator.
- Industrial digital bus communications will eventually make the analog signal output of the integrator an option rather than a standard feature.

---

[4] Rate = Load × Speed

93

- Digital components and the ease of programming software have increased the number of integrator suppliers, many of which are not actually weighing companies.

> **NOTE:** Integrators provided by non-weighing suppliers are often complicated, difficult to use, and operationally deficient.

Basic integrator operation

Siemens Milltronics marketed the first microprocessor based integrator, the Milltronics CompuScale®,[5] integrating the processing power of a computer with the weight calculation of a belt scale system.

Like a computer, the integrator's high speed microprocessor processes input data into digital information. The input data comes in two forms:

- digital square wave input from the speed sensor, itself powered from the integrator
- analog signal from the load cells, also powered from the integrator

The load cells' analog signal is processed by a pre-amplification stage (P/A) before it is changed to digital format by an analog to digital converter (A/D). The microprocessor can do the basic calculations, format, and then display rate, load, speed, and totalization. Data can also be converted back to an analog signal for remote display or alarming.

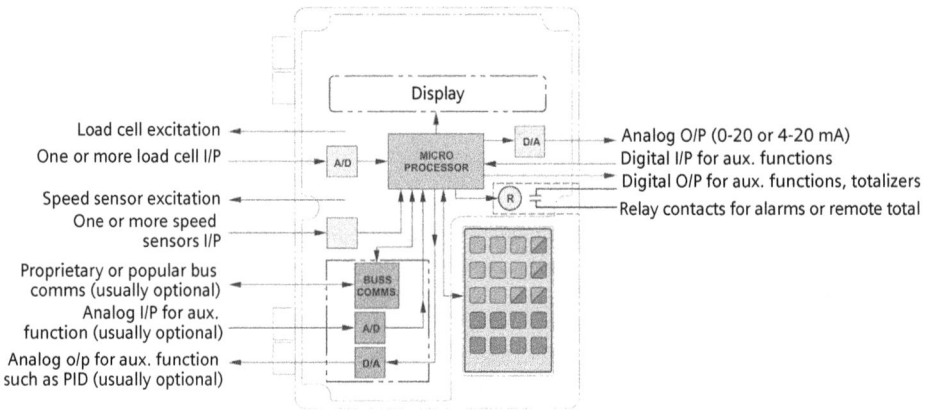

---

[5] *Milltronics CompuScale* is a registered trademark of Siemens Milltronics Process Instruments.

The microprocessor communicates on internal buses, and has a key pad and a digital display for operator programming interface and run-mode monitoring of process values.

Initially, the firmware was stored in Electrically Erasable Read Only Memory (EEROM) chips, and the temporary storage of measured and calculated values occurred in another chip acting as Random Access Memory (RAM). However, the need to store critical data such as application specific parameter information and totalizer values as well as calibration data meant that the RAM had to be backed up by battery power. Battery backup is not always necessary as Non Volatile Random Access Memory (NVRAM) chips are readily available.

Current microprocessors can have non-volatile onboard flash memory, replacing the need for ROM and allowing storage for critical parameters and totalizer values on power down. The flash memory, also available in a separate chip, can be connected to a PC through digital communications so that firmware can be updated and field programmed parameter values can be stored off line and updated. RAM chips are still required for temporary storage because flash memory is limited to a set number of data rewrites.

Basic integrator programming

The integrator needs to be programmed for the application, and while suppliers have different formats for entering the field data, the same general principles apply. Parameters defining the following are common:

- the speed signal – defined in pulses per meter/foot
- the design rate[6] and the design speed[7]
- the calibration reference (chain, weight, or other)
- the value for the calibration reference in terms of loading in kilos/meter or pounds/foot
- the belt length, so the integrator can determine the occurrence of an exact belt revolution, such that zero and span calibrations can be done over one exact belt revolution, or multiples of the same

These and other parameters are usually stored in Flash memory or in battery-backed RAM.

---

[6] 100% value of rate.
[7] 100% value of conveyor belt speed.

Example: primary parameters for configuring the Siemens BW500 Integrator		
P001	Language	English, German, French, or Spanish
P002	Test reference selection	Weight, chain, or electronic
P003	Number of load cells	One, two, or four cells applied to integrator
P004	Rate measurement system	Imperial or metric
P005	Design Rate Units	Imperial: STPH, LTPH, lb/min, and lb/h Metric: t/h, kg/min, and kg/h
P008	Date	Not critical for initial start up, but good to do
P009	Time	Not critical for initial start up, but good to do
P011	Design rate	Full scale capacity of the conveyor (e.g. 300 STPH or 400 t/h)
P014	Design speed	Maximum conveyor belt speed (e.g. 300 fpm or 1.85 m/s)
P015	Speed constant	Value of the speed pulses (e.g. 50 pulse/ft or 150 pulses/m)
P016	Belt length	Measured belt length in feet or meters

Belt scale calibration

Two calibrations are required when setting up the belt scale system:

Zero calibration: adjusts out the tare load (dynamic weigh frame, suspended conveyor idler, and the running conveyor belt), similar to adjusting a bathroom scale for zero without standing on it. The zero calibration is based on the average of the dynamic load cell readings established over one belt revolution or whole multiples of the same. Since the lineal weight of the belt varies over the belt circumference, the signals are averaged digitally while the conveyor is running and the initial zero digital reading is used for future reference (see graphic on page 97).

Subsequent zero calibrations will report percentage error deviation of full scale span, allowing the operator to accept or refuse the correction.

Span calibration: establishes the 100% load value. The span calibration is usually an average digital reading over one, or more, whole belt revolutions with the test reference applied. The load cell signals are averaged digitally and the initial span digital reading is used for future reference.

With the value of the reference, such as with a calibration weight, in terms of kilos per meter or pounds per foot, established during the integrator initial programming stage, the microprocessor calculates the number of counts that is used as the 100% load reading.

NOTES:
- Zero calibrations are required more frequently than span calibrations because conveyor belts vary in weight per lineal length due to temperature and conveyor structural influences.
- Once the initial calibrations have been done, subsequent zero and span calibrations will report percentage error deviation of full scale span giving the operator the option to accept the correction.

Digital Counts: the digital output readings of an analog/digital (A/D) converter rebased on the millivolt input from the load cells. Depending on the pre-amplifier stage and A/D convertor combination used, the number of counts generated by one millivolt signal generally ranges from 1800 to 3600, possibly more. This diagram shows settings typical of Siemens BW500 Integrator with two summed load cell inputs. The number of counts for the zero, test reference, and span are all stored in non-volatile memory for reference once the calibration is complete.

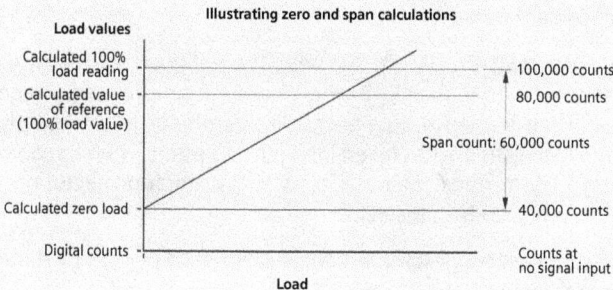

Illustrating zero and span calculations

Load values	
Calculated 100% load reading	100,000 counts
Calculated value of reference (100% load value)	80,000 counts
	Span count: 60,000 counts
Calculated zero load	40,000 counts
Digital counts	Counts at no signal input

Load

NOTE: calculations are performed over one or more belt revolutions

Factoring: with material verification tests, span calibration (and thus the span count) can be factored to account for the influence of the conveyor belt on the actual weighing process. Span calibration accuracy is often affected by two factors:

- the belt's ability to support material
- the weighbridge's efficiency in measuring vertical forces

Most integrators can factor in the span calibration to compensate for these constant impediments and adjust the value of the calibrated reference value accordingly.

Span calibration

There are several methods for setting the belt span calibration reference.

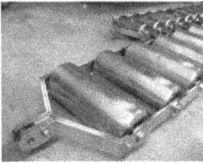

Test chains: provide a calibration reference close to actual because the chain weight is transferred through the belt. Although it cannot duplicate material loading exactly, it provides a reasonable facsimile. These devices can be used when calibrations to material tests are not practical.

> **NOTE:** The placement of the calibration chain on the belt conveyor needs to be repeatable for best results. Material tests and calibration factoring are still necessary for absolute best results.

Static test weights: although they do not duplicate conveyor loading accurately, they do provide a very repeatable reference that tests the integrity of the weighbridge and its load cells each time a span calibration is done. Material tests and calibration factoring are still necessary for best results.

Reference resistor circuit: the resistive network works with the resistance of the load cells in the weighbridge to create a repeatable millivolt reference. However, the integrity of the weighbridge and the load cells are not taken into consideration when a span calibration is performed. Material tests and calibration factoring are still necessary for best results.

Some suppliers suggest a theoretical calibration by using a number of parameters:

- load cell sensitivity
- load cell excitation
- idler spacing
- conveying angle
- design rate and design speed to calculate the exact mV output of the load cells and thus the digital span count expected at the design rate

The span calibration can be done without running the belt, or applying a dynamic or static weight as reference. Only a zero calibration has to be done at the time of commissioning the belt scale. Like the other calibration reference options, material tests and calibration factoring are still necessary for best results.

> **NOTE:** Measurement Canada, NTEP, OIML, and MID certified belt scale systems do not allow for span calibration once the calibration accuracy is verified and factored by material tests. Span calibration is not considered necessary with current approved system components that hold the span calibration indefinitely. Zero calibration is done to compensate for deviation in belt weight caused by temperature variations.

A customer was using an MMI belt scale system to obtain Measurement Canada certification. Unfortunately, when testing for certification, it was slightly out of the allowed deviation from 20% to 100% of the conveyor capacity. The installation was mechanically pristine, but the load cell output wires were all connected in parallel, keeping the load cell balancing feature of the CompuScale III from being used.

By rewiring the load cells to use the differential load cell inputs with software based load cell balancing, the Siemens technician achieved a 0.25% accuracy improvement. As a result, the belt scale system easily passed the next Measurement Canada inspection, thereby allowing it to be used for custody transfer.

Integrator designs

The stand alone integrator for one belt scale system is the most popular design on the market. Some manufacturers also offer wall mounted and panel mounted versions as well as PLC-based weighing modules.

Stand alone

The most common integrators are designed for mounting close to the weighbridge and speed sensor with packaging to meet in-plant locations.

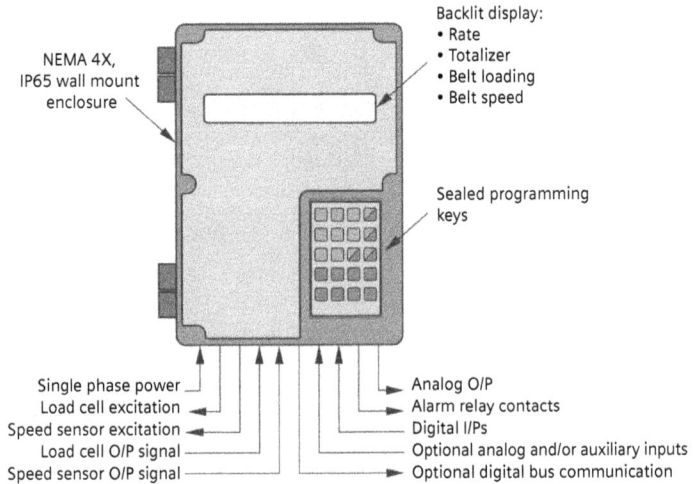

NEMA 4X,
IP65 wall mount
enclosure

Backlit display:
• Rate
• Totalizer
• Belt loading
• Belt speed

Sealed programming
keys

Single phase power
Load cell excitation
Speed sensor excitation
Load cell O/P signal
Speed sensor O/P signal

Analog O/P
Alarm relay contacts
Digital I/Ps
Optional analog and/or auxiliary inputs
Optional digital bus communication

PLC based

Users with existing or planned Siemens PLCs operating in their plant can avail themselves of Siemens' SIWAREX weighing modules like the SIWAREX FTC, specifically designed to work with a belt scale system or weighfeeder as a field device. These modules provide high precision measurement and computations with a limited burden on the PLC supervisory system.

Integration programmed into a PLC operating system

This method has had limited success. Programming time is expensive when only a few applications are required. The integration function is also taxing on the PLC's CPU time. Eventually, most users take the integration function outside of the PLC or DSC operating system and rely on the expertise of the belt scale system component supplier.

Seamless integration to PLC operating system
• operation power from PLC power bus
• display on PLC HMI

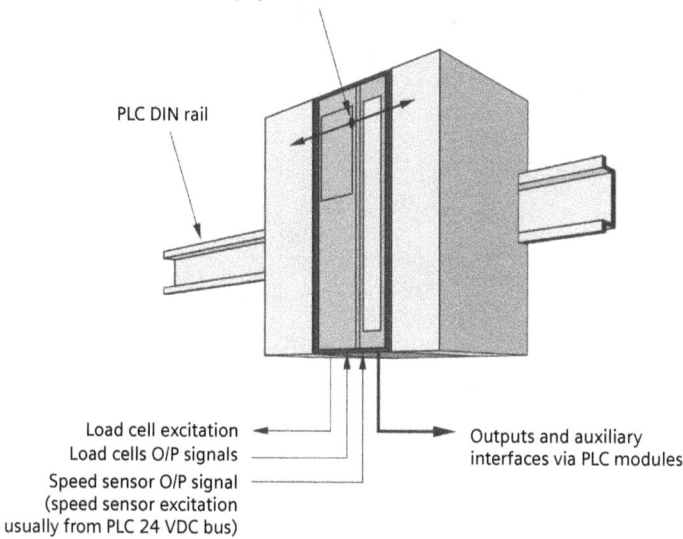

PLC DIN rail

Load cell excitation
Load cells O/P signals
Speed sensor O/P signal
(speed sensor excitation
usually from PLC 24 VDC bus)

Outputs and auxiliary
interfaces via PLC modules

Multiple integrators

For multiple belt scale system integrators, processor speed presents a problem because the many, continuous, rapid calculations require a fast microprocessor. Another configuration requires a separate microprocessor for each belt scale system with a master microprocessor accumulating data and acting as the interface controller.

A more viable option uses Siemens SIWAREX FTC modules with a Siemens PCS7 control system. The power supply, analog o/p, and/or bus communications are controlled by the host operation system, but the integration of each belt scale system is handled by an FTC module.

101

Optional display

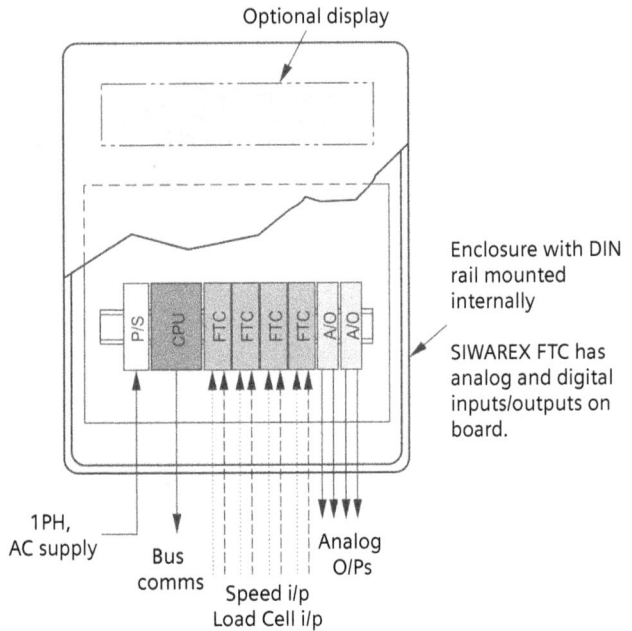

Enclosure with DIN rail mounted internally

SIWAREX FTC has analog and digital inputs/outputs on board.

1 PH, AC supply

Bus comms

Speed i/p

Load Cell i/p

Analog O/Ps

NOTE: A dedicated Siemens PLC along with Siemens SIWAREX FTC modules and other functional modules can be used as a multiple belt scale integrator.

Load cell signal processing

Signal processing does not vary significantly among integrator designs. However, Siemens Milltronics has differentiated itself by greatly improved processing through the use of differential load cell inputs.

Traditional single load cell input

The most common input circuit into an integrator is a single analog input applied through a pre-amplifier and an A/D converter to the microprocessor. With more than one load cell, the outputs must be all wired in parallel. Matching the load cells for sensitivity is necessary, and special attention needs to be paid to cable lengths to ensure equal resistance for the load cell circuits. Each of the parallel circuit resistances must be matched to the others, usually done with tuning resistors and in a summation junction box circuit.

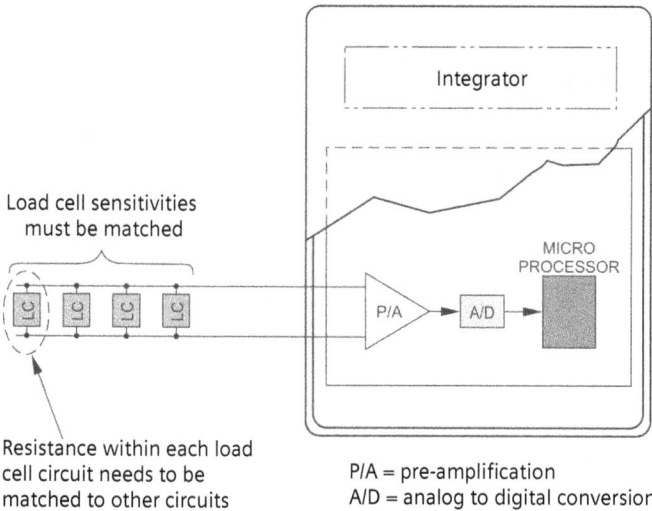

Load cell sensitivities
must be matched

Resistance within each load
cell circuit needs to be
matched to other circuits

P/A = pre-amplification
A/D = analog to digital conversion

Analog load cell interface (one, two, and four load cell system)

With an individual pre-amplifier (P/A) and A/D convertor for each load cell, the load cells can be applied to the integrator in a *differential* manner. The micro-processor sees one or two distinct load inputs in digital format and adds them in the calculation of the digital reading. Each A/D convertor is linked to the common load cell excitation and adjusts its gain inversely proportional to any changes in the excitation supply level.

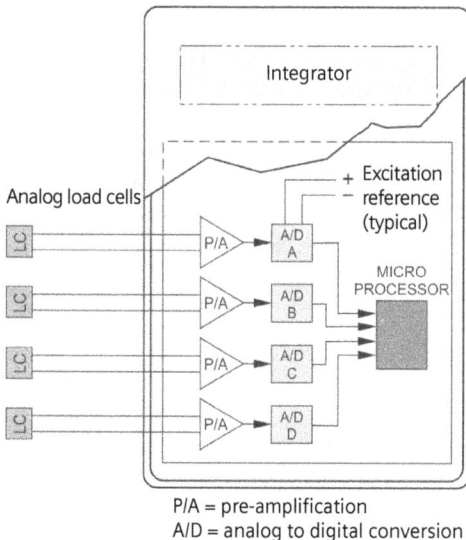

P/A = pre-amplification
A/D = analog to digital conversion

103

Siemens BW500/L

- one load cell: only the output of A/D convertor A is used by the microprocessor
- two load cells: the output of A/D convertors A and B is used by the microprocessor with the total reading being a sum of the two A/D outputs

Total possible span resolution for a two load cell system is approximately one in 500,000.

Siemens BW500 (four load cell inputs)

- when four load cells are used, the output of the sum of A/D convertors A and B are added to the sum of A/D convertors C and D
- load cells A and B and the load cells C and D can be balanced by the software to compensate for any mismatch in sensitivity or circuit resistance
- total span resolution for a four load cell system is approximately one in 1,000,000

Digital load cell interface[8]

By having the pre-amplification (P/A) and A/D conversion, as well as a small microprocessor in each load cell, the load cells can be applied to the integrator via bus communications. Therefore, the integrator's microprocessor sees a number of distinct digital load inputs and adds them into the calculation of the digital reading. A lot of the digital interface is still in development, but there are numerous benefits:

- circuit resistance within the inter wiring is not a factor
- each load cell is seen as an individual device and is read, in turn, by the microprocessor
- digital counts from each load cell provide data in measurable quantities (kilos or pounds). The data permits easier balancing of the load cells and simpler integration of the load signal and the speed sensor signal.

---

[8] During publishing a new PLC based weighing module is being developed. The SIWAREX WP241 is a new belt scale module in the S7-1200 PLC family. The module features a new stand alone functionality so that it can be used with any PLC system or directly with an HMI. The module has four digital inputs and four digital outputs as well as a 4-20mA output. More importantly it has modbus TCP I/P and RS485 communications as standard.

- belt speed sensors will eventually become smart sensors and connect with the integrator via the same data highway as the load cells
- a multiple belt scale integrator using a high speed micropro-cessor is more feasible

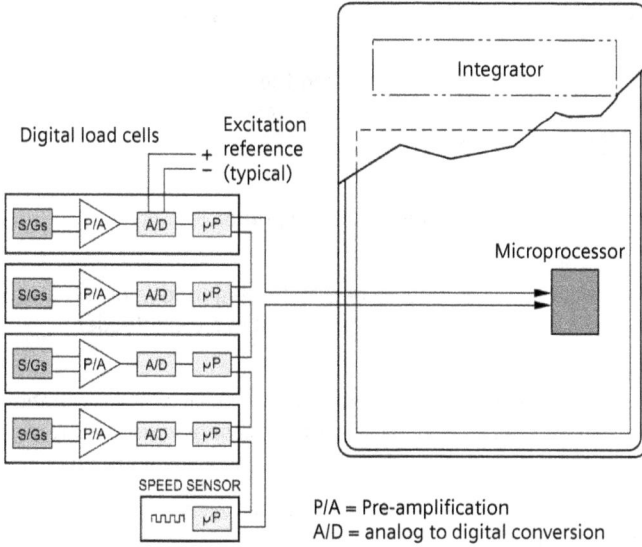

Eventually the connection from the belt scale and speed sensor to the integrator will have a wireless option. Wireless communication between the integrator and the plant control system will also be the norm.

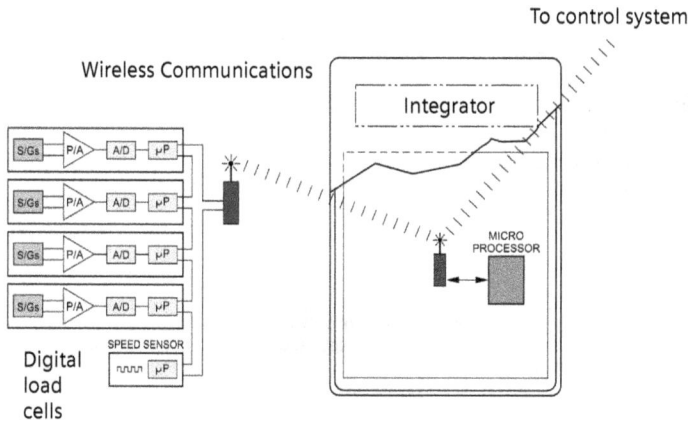

105

Common integrator features

Automatic self initiated zero:
- the integrator senses an empty belt by monitoring load cell signals and automatically initiates a zero calibration
- the deviation will be automatically adjusted, as long as it falls within pre-set boundaries
- totalization will not be lost if the conveyor loading is dramatically increased
- an effective feature for when the conveyor does not have a gravity take-up. It ensures best consistent belt tension and/or good belt tracking control

Communications ports:
- for direct connection to a serial printer

PID control:
- generally used when a belt scale weighbridge or weigh deck functions as the weighing element in a weighfeeder

Wireless interfaces:
- increasing in popularity, especially in outdoor site applications
- please note that wireless communications must be supported by both effective scale design and quality raw signal processing

Bus communications – at least one of the following:
- Modbus RTU
- Profibus DP
- A-B RIO[®9]
- DeviceNet[™10]

Ethernet connectivity:
- for monitoring by a PC
- for a plant control system

Bus communications (eg. Siemens SITRANS RD500[11]):
- process monitoring
- inventory data monitoring
- alarm upset conditions

---

[9] A-B (Allen-Bradley) is a rgistered trademark of Rockwell Automation.
[10] DeviceNet is a trademark of Open DeviceNet Vendors Association (ODNA).
[11] Remote Data Manager

Conveyor slope change management:
- set programmable parameter for degrees of incline (or decline) to compensate for changes to gravitational forces that are the result of changes in conveyor incline. This affects the load reading and has a bearing on the rate and total calculation
- set the microprocessor to read the analog signal input from a inclinometer so the load calculation automatically adjusts with the changing slope, changing the rate and total readings
- adjust the load cell excitation voltage to compensate for load cell output changes due to the changes in incline or decline

Dry rate (theoretical determination):
- programmable moisture parameter for moisture percentage to calculate dry weight, affecting the load reading and thus the rate and total calculation
- use a moisture meter to provide an analog signal to the microprocessor so the load calculation, and thus the rate and total readings, is automatically adjusted

Differential speed detection:
- second speed input so the speed rate can be detected at two separate locations
- second input helps detect upset conditions such as drive pulley slippage and belt breakage
- useful in weighfeeder applications

Multi-span and multi-zero features (for calibrating the system in any of these conditions):
- reversing conveyors (multi-zero and multi-span)
- conveyor with more than one feed point and/or more than one material

Linearizer function:
- compensates for load reading inaccuracy, occasionally a problem at the low end of operational loading (usually below 25%)

Recording control system totals

Although many integrators offer bus communications that permit a central control system to read the totalizers, the greatest accuracy is still found with the remote totalization function of the integrator itself. The control system will be wrong when the analog rate output signal is applied as a process indication and then integrated to derive a running total. There are three reasons the control system will be incorrect:

- the analog signal and the actual rate do not match (is 4.00 mA really 0.00% rate / is 100% rate really 20.00 mA?)
- the PLC will not have totalizer drop out, so it can continue to total while integrators generally do not do so at near zero rate
- negative totalization is not possible in the PLC. An integrator will totalize both positively and negatively as the belt is calibrated and the reference percentage is set

For a control system to operate as accurately as an integrator, it requires two or more conductors beyond the analog signal conductors so the integrator totalization is accurately recorded.

Interconnection of belt scale system components

Follow the interconnecting wiring guidelines published by the integrator supplier. Incorrect cable and conductor sizes, and improper grounding techniques, can cause start-up and operational problems. Use the correct shielded multi-conductor cable and the right conductor size and number for system success.

Comparison of integration options

Function	PLC computation software	PLC belt scale integrator module	Stand alone integrator
Load cell excitation	By PLC power module	Included	Included
Speed sensor excitation	From PLC bus, or power module	From PLC bus, or power module	Included
A/D load cell signal conversion	PLC analog input module	High resolution included	High resolution included
Speed signal reading	PLC digital input module	Included	Included
Integration calculations	Within PLC operational system	Included	Included
Display	Engineered HMI (sample routines available for programmer)	Engineered HMI (sample routines available for programmer)	Included
Zero calibration	Within PLC operational system (functionality to be programmed)	Included	Included
Span calibrations	Within PLC operational system (functionality to be programmed)	Included	Included
Remote totalization	Not necessary	Direct transfer to PLC operating system	Solid state relay (to PLC or via industrial communication system)
Analog output	Only necessary when going to a field device when analog module is required	Standard	Standard
PID control	Within PLC operational system	Within PLC operational system	Optional
Demand analog control signal (PID)	PLC analog output module	Onboard analog output can be used	Optional (functionality for weighing applications embedded in firmware)
Alarm relays	PLC contact output module	PLC contact output module	Included (functionality for weighing applications embedded in firmware)
Fieldbus communication	Host PLC bus only	Host PLC bus only	Optional: Modbus, Profibus, DeviceNet, etc.
Printer	PLC system printer (functionality to be programmed)	Serial printer can be directly connected to onboard RS232 interface	Optional to serial printer (functionality for weighing applications embedded in firmware)
Summary	Host computer slows down as PLC needs to do all calculations	Belt scale integration calculations done in module only	Belt scale integration calculations at integrator only (user interface embedded in package)

Summary

The integrator is where all the action comes together for the weighing system. The load cells and the speed sensors are busy feeding the integrator all the bits of data required to give the necessary weight reading. Whether it is the stand alone Siemens BW500 or an integrated SIWAREX FTC linked into a Siemens central control system, the confluence of data must be correctly managed for accuracy and repeatability. Thus calibration must be accurate, installation must be precise, and programming must take all the application variables into account.

Chapter 8
Belt Scale Installation

It's not the teaching, it's the learning.[1]

The best scale design, load cells, speeds sensor, and integrator mean little if the system is incorrectly installed. This chapter outlines the best installation practices to ensure that maximum value is realized from the belt scale application.

Topics

- Terminology
- Installation factors
- Idlers
- Installation guidelines
- Maintenance and modification

Belt scale/conveyor terminology

Aside from the electronic parts, a belt scale conveyer system has numerous components whose operation can affect the quality of the weight measurement.[2]

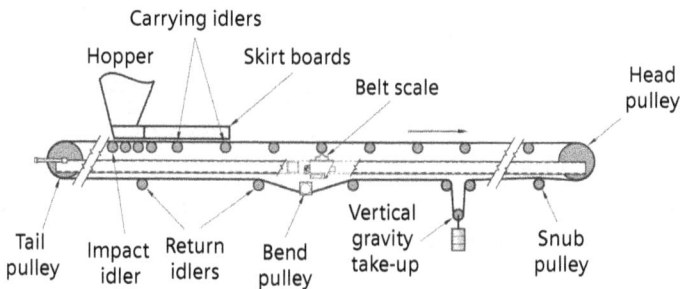

Carrying idlers
Hopper
Skirt boards
Belt scale
Head pulley
Tail pulley
Impact idler
Return idlers
Bend pulley
Vertical gravity take-up
Snub pulley

Scale area idlers

The scale area idlers are designated according to location and belt travel:

---

[1] Sly and the Family Stone, "In Time." *In Time*. 1973.
[2] This diagram labels the various conveyor parts for easy reference. The precise definitions can be found in the glossary (See *Appendix A: Glossary*).

- Approach (A) – A1 is always last idler before the scale idler S1
- Scale (S) – where the measurement happens
- Retreat (R) – R1 is always the first idler past the scale idler S1

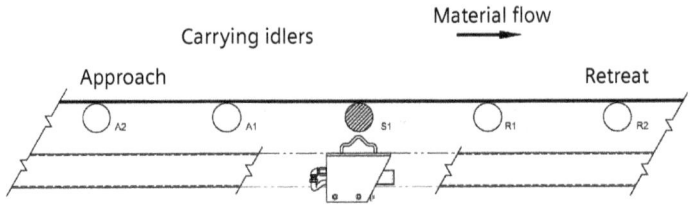

NOTE: A second scale supported idler would be S2, located between the S1 and R1.

Installation

Successful belt scale installation and optimal performance depends on these four criteria:

1. Belt scale location on the conveyor
2. Mechanical and electrical components
3. Idlers
4. Maintenance and future modifications

NOTE: All diagrams are specific to a single idler belt scale. When applying the recommendations to other belt scale models, use the diagrams as examples only.

Belt scale location

The placement of the scale in the conveyor must be precise and account for the following:

- belt tension
- material turbulence
- curved conveyers

Belt tension

Belt tension varies in relation to material tonnage, belt speed, conveyor length, and the height that the material must be raised. The greater these system values, the greater the tension on the belt, and the increased tension affects the scale's accuracy. When the belt tension is high, the conveyed load transfers with minimal belt deflection, reducing accuracy potential. Furthermore, a slight idler misalignment when the belt operates under high tension can lead to non-linear weighing results over the loading range.

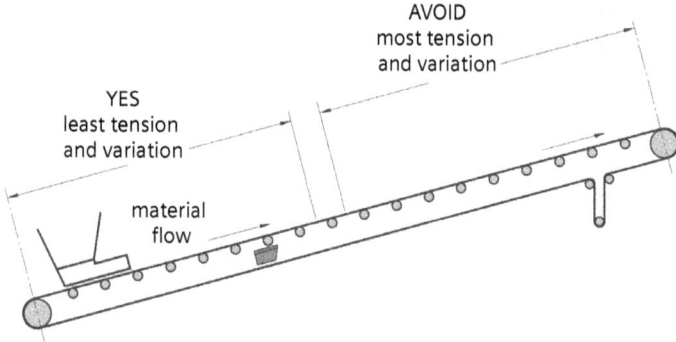

Install the scale close to the tail section where tension and tension variations from no-load to full-load are minimal.

Material turbulence

Material leaving the feed point area and skirt boards will be turbulent, requiring some belt travel to settle. Do not weigh the material before it settles completely.

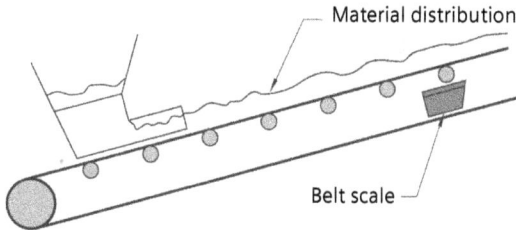

Recommendation: Locate the scale no less than one idler space beyond the point where turbulence stops. If that cannot be determined, use these recommendations for scale location.

NOTE: Suggested distance from the end of the skirt boards to the scale based upon belt speed.	
Belt speed	**(d) Distance**
up to 1.5 m/sec (300 fpm)	2 m (6 ft)
up to 2.5 m/sec (500 fpm)	3 m (10 ft)
over 2.5 m/sec (500 fpm)	5 m (15 ft)

113

Curved conveyors

Vertical conveyor curvature can have a negative effect on belt scale performance. Both concave and convex curvatures will disturb the idler alignment if the belt scale is installed in the area of the curve. Note that the concave curvature may lift an empty belt from the idlers around the curve, preventing a good empty belt zero balance for the scale.

Minimum distance the belt scale should be from the curvature:

Concave

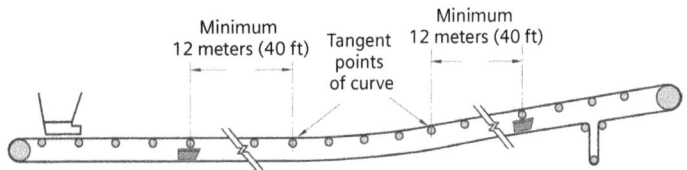

Convex

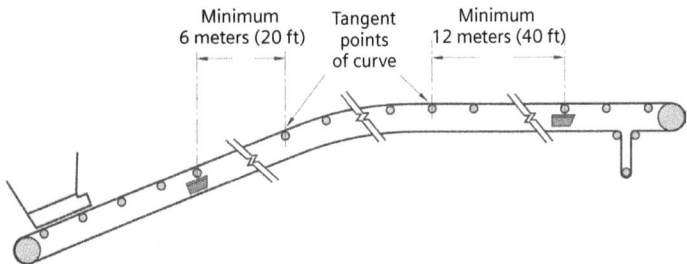

Conveyer types

Belt conveyor diversion ploughs

Belt conveyor diversion ploughs grade the material off to the side of the belt. Do not place a diversion plough, or any other conveyor or material control device that changes the profile of the carrying belt, near the scale installation. These devices alter the belt scale idler alignment and may also create drag on the belt which the scale reads as material force or load. Do not install the belt scale within nine meters (30 feet) of belt ploughs or profile changing devices.[3]

---

[3] For information on diversion ploughs see: http://www.handbmining.com.au/2008/10/31/carry-belt-diversion-plough/

Stacker conveyors

Any conveyor that is not a permanent structure or that varies in its incline, elevation, or profile is not considered a good installation for an accurate belt scale. For applications where a belt scale can be used effectively with a stacker conveyor, please consult your Siemens weighing representative.

Conveyor trippers

A conveyor with a tripper discharge is not as common as a conveyor with fixed curvature, but it can have a similar effect on belt scales. A tripper can cause varying belt tension and can lift the belt in the scale area if the belt scale is not located in the best position. On a conveyor with a tripper, locate the scale according to the recommendations for fixed curves, but with the tripper in its fully retracted position.

A tripper discharges material from a belt.

Other installation considerations

Belt scale take-up device

Maintaining proper belt tension is crucial and there are three basic conveyor belt take-ups to control conveyor belt tension:

- screw: limited to conveyors with pulley centers of less than 18 meters (60 feet)
- vertical gravity: the vertical gravity take-up is the most effective because it maintains relatively uniform belt tension, reducing the influence of belt tension on the scale and improving accuracy
- horizontal gravity: only if vertical gravity take-up is not practical nor possible

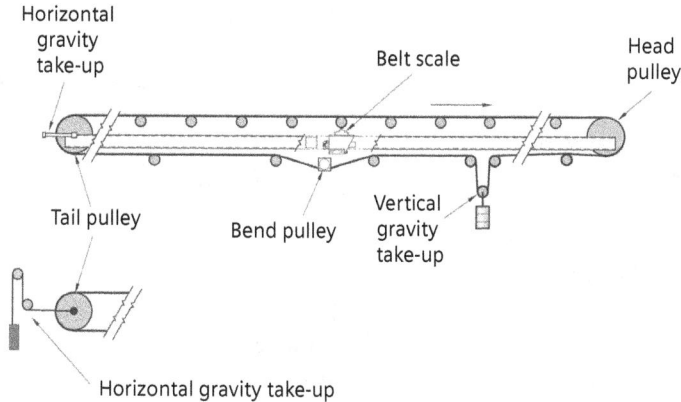

Material feed point

Multiple feed points can change belt tension considerably, depending upon the combination of feed points used. Whenever possible, install the scale on a conveyor that has only one feed point.

Material loading

The flow of material from the pre-feeding device to the belt is often not uniform and occurs at speeds different from the conveyor belt, reducing accuracy. To ensure steady and uniform material loading to the conveyor belt at or near the same speed as the belt, install a material feed control gate or similar device to deliver uniform material flow.

With control gate

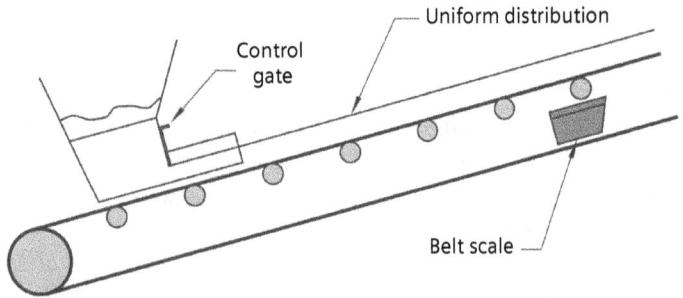

Without control gate

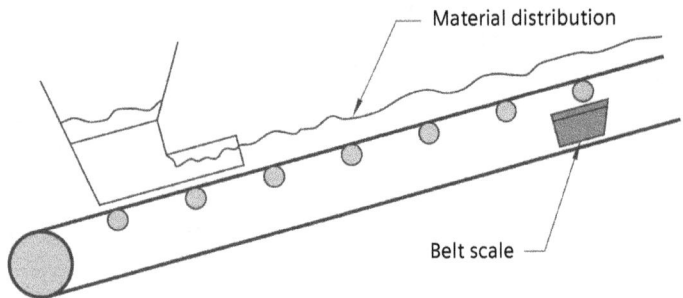

Material roll back

Material roll back (a.k.a. material slip) occurs when the belt material moves back on itself. Material moving against the motion of the belt disrupts the accurate weighing process. The following factors can contribute to roll back:

- material type – granular spheres roll more easily than flake type material
- conveyor with a steep incline
- inequality between material feed velocity and belt speed
- poorly selected or badly installed rubber or chain curtains at the in-feed, where the curtain momentarily holds back the material on the top of the pile causing it to slow down in comparison with the rest of the pile

---

NOTE: Make sure to inspect the installation closely for proper speed-to-feed and incline.

Conveyor belting

Belts are the snowflakes of weighing technology in that they are rarely alike. The variations in the number of belt plies, the cover thickness, and the type and quantity of splices in a belt cause considerable deviation in weight throughout its length. During zero balancing most belt scales average the weight of the belt over one complete circuit and any deviation (+ or –) from that average may make it difficult to obtain a good zero reference, subsequently affecting scale accuracy.

NOTES:
- A belt that is over rated for its intended use may be so stiff that it cannot flex enough to properly trough in the idlers (especially in 35° and 45° idlers). The belt then arches across the idler and neither a good zero calibration nor a good span calibration are obtained.
- When replacing worn sections of belting, ensure that they are the same as the existing belting.
- When choosing a new belt, select one that suits the application. Avoid selecting an over rated belt.

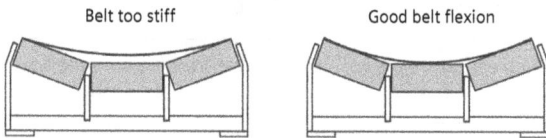

Belt too stiff   Good belt flexion

Belt tracking and troughing

Proper tracking of a conveyor belt means that the belt keeps its position on the conveyor and idler center line. It must also lie evenly in the idler trough and make good contact with all three idler rolls. Proper tracking requires the belt to be right for the conveyer, the conveyer to be right for the application, and the idlers to be installed properly.

belt:
- sufficient ply rating to support the load without being over rated
- rubber covers are of the proper thickness
- splices are properly selected and installed

conveyer:
- the conveyor take-up must be the right type for the application, adjusted and working properly

idlers:
- idlers square to the conveyor and located centrally on the frame
- all idler rolls turn on their axis
- training idlers or idlers with guide rollers are not installed closer than nine meters (30 feet) from a scale idler

Idlers

There are several basic idler maintenance principles that will keep the installation running accurately and reliably:

- employ idlers designed specifically for belt scales to ensure good idler alignment in and around the scale area
- keep all idler rolls clean, free from material buildup, and spinning free without over greasing
- replace all idlers that are stiff, stopped, or have eccentric rolls
- all idlers chosen for scale installation must be of the same manufacture and properly lubricated. Some installations may require idlers with lube-for-life bearings.

NOTE: Neglecting simple maintenance can result in misalignment and poor belt tracking.

Basic idler design: the two most effective designs are the troughed three roll in-line idlers or single-roll flat idlers. Do not use wire rope, two-roll V, or cantenary idlers on or near the scale.

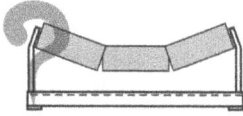

Offset conveyer idler
(acceptable in some applications)

Wire rope idler

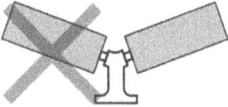

V-roll (two roll) conveyer idler

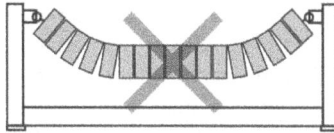

Cantenary idler

Flat idler

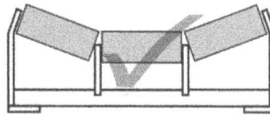

Equal roll troughed idler

NOTE: Offset idlers may be acceptable in some installations (consult your Siemens Milltronics representative regarding their use).

Idler trough angle

- the most commonly used are 20° and 35° troughed idlers
- 45° troughed idlers can be used but accuracy may suffer because the deep trough angle amplifies the effects that belt tension and stiffness have on the scale, increasing the need for proper idler alignment

Idler dimensions

- ensure the idlers are of the same dimension and have rolls that are concentric within 0.5 millimeters (0.02 inches)
- the troughs must match within three millimeters (0.12 inches) when compared to a template

119

Idler alignment in the scale area

Idler alignment is crucial to scale performance and the belt scale manufacturer's guideline must be closely followed:

- align and level the weighing area idlers properly by shimming the scale idler, the two approach idlers, and the two retreat idlers, until they are within ± 0.8 millimeters (0.031 inches) of each other. Make sure they are centered and squared to the conveyor during the shimming process. In high accuracy applications, alignment should be extended to three approach and three retreat idlers.
- precise idler alignment is required for maximum accuracy. Misaligned idlers result in unwanted forces affecting each idler in the weighing area, leading to calibration and measurement errors.

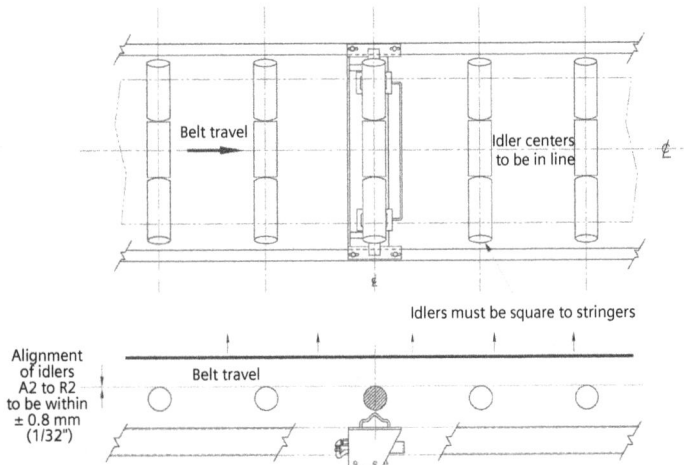

- be sure to use good quality wire or string to check for alignment because it should withstand sufficient tension to eliminate any sag. Adjust shims so that all rolls of the A2 through to the R2 idlers are in line within ± 0.8 millimeters (0.031 inches).

A2
A1
S1
MSI with
modified idler
R1
R2
Wire or string
(for alignment)

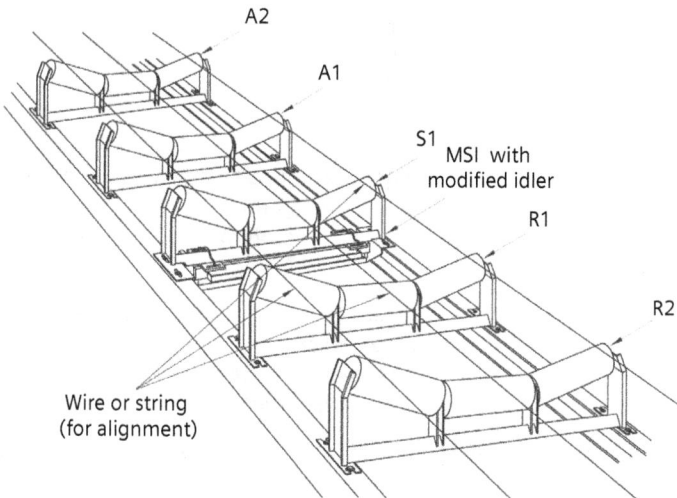

- Although the accepted tolerance for idler alignment is ± 0.8 millimeters (0.031 inches), the scale mounted idler should never be lower than the adjacent idlers. Establishing good idler alignment is the most important part of the installation procedure. Belt scale system accuracy is directly affected by idler alignment.

Belt conveyor idlers in the scale area are not always required to be of "scale quality," so roller run-out specifications may be far above normal idlers. Some belt scale system suppliers may take advantage of the opportunity to make the extra sale with such items even if the specifications are not required.

If idlers are within the quality specifications of the supplier and the supplier makes quality products then they are generally suitable. When cost is no object and the accuracy desired is the absolute highest (0.125 to 0.25%), only then are scale quality idlers recommended.

121

Pulleys

The head and tail pulleys are at either end of the conveyer. The head pulley is at the discharge end of the conveyer while the tail pulley is at the loading end.

Head pulley

Head pulleys are generally flat faced with a slight crown, and when used with troughed idlers, the belt profile must change from troughed to flat. Conveyer designs accommodate this change in several ways:

- by placing the head pulley higher than the top of the center roll of the nearest adjacent idler
- by inserting idlers of decreasing trough angles between the head pulley and the normal idler run

When installing a belt scale in a short conveyor, or if the scale must be located near the head pulley, make sure design accommodations are made to account for the considerable amount of stress exerted on the belt edges and the idlers adjacent to the head pulley. These stress forces are subsequently applied to the scale and have a negative affect on the measurement.

Belt scale placement

A. On conveyors with 20° trough idlers, place a minimum of two fixed 20° idlers between the belt scale and the head pulley.

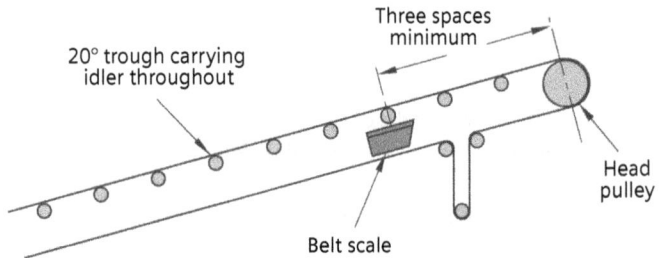

B. On conveyors with 35° trough idlers, place a minimum of two 35° and one 20° retreat idlers between the scale and the head pulley.

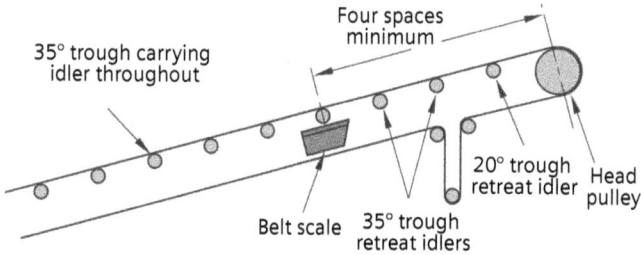

C. On conveyors with 45° trough idlers, a minimum of two 45°, one 35°, and one 20° retreat idlers must be located between the scale and the head pulley.

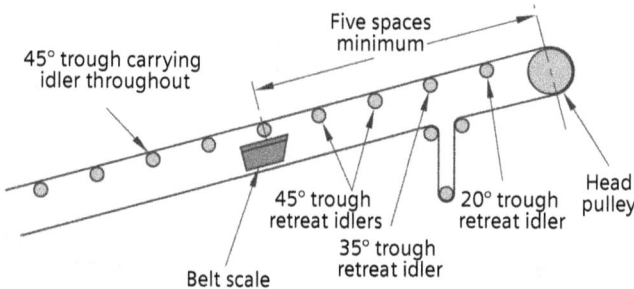

D. The vertical displacement of the head pulley relative to the adjacent retreat idler is normally in excess of what is acceptable for belt scale installations.

When locating a scale close to the head pulley, maintain a maximum of 13 millimeter (0.5 inches) vertical displacement between the top of the head pulley and the top of the center roll of the adjacent idler by making the following adjustments:

- Lower the head pulley on its mounting until the vertical displacement measured from the top of the head pulley does not exceed 13 millimeters (0.5 inches) above the top of the center roll of the adjacent idler.
- Shim all the retreat idlers between the head pulley and the scale, the scale idlers, and at least two approach idlers to achieve the same conditions illustrated in placement option A above.

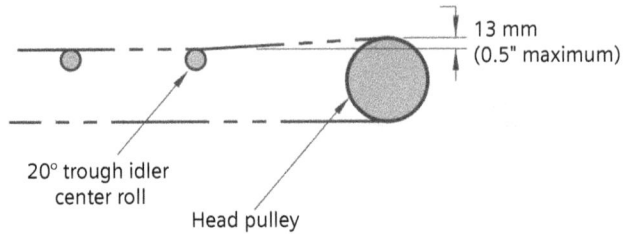

13 mm
(0.5" maximum)

20° trough idler
center roll

Head pulley

Tail pulley

Self-cleaning pulleys are often called 'wing pulleys.

Usually, the space reserved for the in-feed suppresses any effect the tail pulley might have on the scale. A problem could occur if the tail pulley is the self-cleaning type with slats or beater paddles. The beating action of this pulley may create oscillations that could be transmitted through the belt to the scale. If possible, avoid wing type pulleys and use solid face welded steel pulleys.

Conveyor

Conveyor rigidity

The conveyor stringers in the scale area should be strong enough to limit relative deflection to 1.6 millimeters (0.0625 inches) or less with supports 2.4 meters (8 feet) apart throughout the range of conveyor loading. Stringers should also be straight so that the belt has a better chance of tracking centrally on the conveyor.

Conveyor covers

Covers are required for outdoor installations involving belt scales. Ensure that the covers do not interfere with the operation of the scale. Install additional shielding to counteract the adverse effects of the elements (wind in particular). The amount of shielding will depend on the geographical area, but typical dimensions are nine meters (30 feet) before and after the center of the scale and 1-1.2 meters (3-4 feet) above and below the carrying belt line.

Skirtboards and sealing strips

Some applications must extend the in-feed skirtboards and sealing strips to the full length of the conveyor. However, weighing accuracy can be negatively affected by the force the sealing strips exert directly on the belt and indirectly on the idlers, especially when pinching occurs. Obtaining accurate zero balance and span calibrations under these circumstances is difficult.

Belt | Sealing strip | Skirt board

Idler

Belt scale

Pinching may
occur here

Skirt board
Sealing strip may
be dragging along
the belt
Belt

Idler

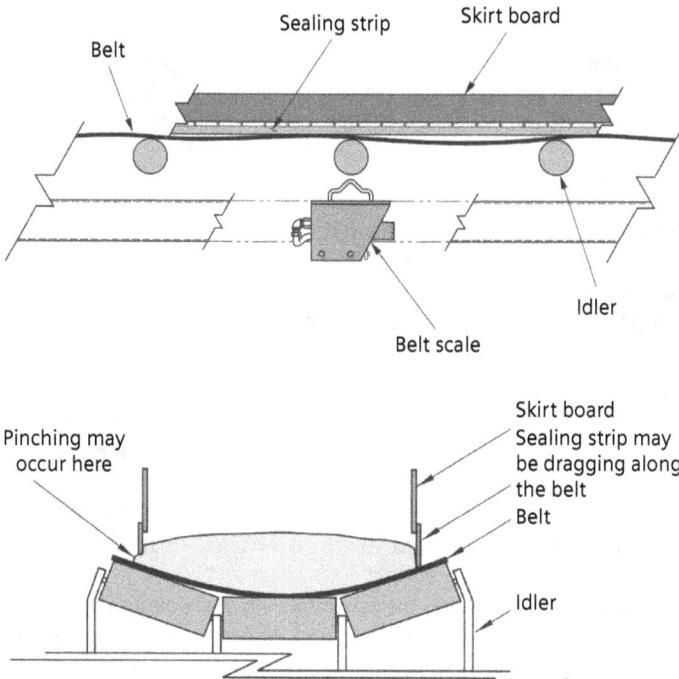

Recommendation: remove the sealing strips or raise them sufficiently to eliminate their effect upon the belt and idlers.

General installation guidelines

Speed sensor installation

The speed sensor must be property mounted, as per manufacturer's recommendations, and ensure the following:

- the mounting puts the lowest level of stress on the wheel or shaft bearings as possible, extending operating life
- a shaft mounted speed sensor needs to be prevented from any counter rotation by a spring or rubber cord
- the body of a shaft mounted speed sensor can only be fastened solidly to the conveyor structure if there is a rubber flex connection connecting the shaft to the drive shaft of the belt driven pulley

125

Integrator installation

The integrator is best installed near the weighbridge to make it easier to set up, calibrate, and provide local displays. Non-hazardous environments makes this possible since most self standing integrators have IP65 ratings[4] and are suitable for even dusty and corrosive environments.

In hazardous applications, the integrator must be installed in a non-hazardous area, ensuring that the belt scale and the speed sensor have the necessary approvals. Make sure that the belt scale system components are suitably rated for the application environment.

Connecting components

Please refer to user manuals for all specific installation information.

- The integrator, weighbridge, and speed sensor must be connected with shielded multi-conductor cable as per the integrator manufacturer's recommendations.
- Speed sensor: connect both two or three conductor cable versions to the integrator with 18 AWG (0.75 mm²) shielded cable
- Load cells: connect to the integrator with the minimum of 18 AWG (0.75 mm²) shielded cable and use a greater core size for long runs. Also, for long runs, two extra conductors can be used to sense the excitation voltage at the load cells. These sensing conductors allow the integrator to measure and adjust for voltage drops across the excitation conductors to maintain consistent excitation.
- Integrator: any analog signals (usually 4-20 mA) to and from the integrator to other devices such as a PLC or DCS control system requires a minimum of 18 AWG (0.75 mm²) shielded cable. Bus communication connections must be compliant with the standard practices.

WARNING NOTES:
- For hazardous zone applications, make sure that the cabling and/or conduit requirements are met.
- All electrical cable shields must be connected to ground (earth) at the integrator and not at the load cells, or speed sensor, or at the control system end.
- Never ground the shield of a shielded cable at both ends in order to avoid ground loops.

---

[4] NEMA 4, NEMA 4X

Maintenance and modifications

Maintenance

Once the conveyor is fitted with a belt scale, it requires more attention as it is now part of the weighing system. To ensure accurate weighing, be sure to do the following:

- lubricate all pulleys and idlers
- ensure proper belt tracking
- clean and scrape belt properly
- operate with proper belt take-up
- design for proper material and spillage control
- train personnel on maintenance and safety

WARNING NOTES:
- When welding near the belt scale, ensure no current passes through it.
- Reset the shipping stops during maintenance to reduce physical shock to the load cells.
- Avoid vibration and impact on the scale. A belt scale should be isolated from equipment that can induce harmful or disturbing vibration. Equipment such as crushers, vibratory feeding equipment, bins subject to hammering, and hammer mills should be avoided.

Modifications

Any changes to the conveyor and/or related equipment could have a dramatic effect upon the operation and resulting accuracy of the belt scale. Always consult your belt scale system supplier representative for advice regarding belt scale installation in a modified conveyor system.

Material buildup

Keep the conveyor belt and associated equipment as clean as possible, so that the scale measures only the loads intended and not the added weight from material stuck to the belt.

Remove sticking materials with quality belt cleaning equipment:

- belt scrapers
- rotary brushes
- vibrating cleaners
- shakers
- ploughs

Material spills

General good housekeeping is always important. Material spillage leads to lost production and affects scale operation when overflow material wedges between dynamic and static parts preventing proper scale deflection. Buildup also affects the zero balance of the scale.

Do not overload the conveyor. As a precaution, install deflectors to keep spills from reaching the scale area.

Summary

Setting up an accurate and dependable weighing system depends on getting the right combination of conveyor conditions, measuring technology, and careful installation practices. Paying attention to these details will ensure that the system provides clean data to the integrator so that it can calculate the moving material accurately and provide dependable control to the operator.

However, like any professional, the system is only as good as its tools. And the Siemens belt scale is the best in the business. A properly installed Siemens Milltronics MSI and the Siemens Milltronics BW500 provide years of dependable and accurate weighing data. The next chapter gives examples of some of the many installations that successfully use Siemens weighing equipment in applications ranging from cement and aggregates to dog food.

Chapter 9

Applications and Summary

Flying is learning how to throw yourself at the ground and miss.[1]

Siemens weighing technology represents the culmination of decades of personal experience of our technicians, application engineers, and customers. Weighing is a rugged business and we don't mind getting our hands dirty, providing quality dynamic and static weighing devices that perform accurately and reliably.

With state of the art integrators and PLC based weighing modules working in tandem with rugged and reliable belt speed sensors, Siemens belt scales (especially the MSI) have proven themselves to be the best on the market.[2]

Siemens weighing equipment is particularly suited for mining, aggregates, and cement applications. However, where ever material and product are being moved, from dog food to headache pills, belt scales, weighfeeders, and solids flowmeters play an important role in calculating weight.

Topics

- Application overview
- Cement
- Mining
- Aggregates
- Food and beverage
- Applications in general industry

Application overview

Mining and aggregates

Belt scale applications are strong in mining in numerous functions:

- tabulate the conveyed run-of-mine (ROM) exiting the mine
- monitor the input and output of crushing mills

---

[1] Douglas Adams, English writer. 1951-2001.
[2] For more information on Siemens weighing, go to www.siemens.com/weighing. Resources like the *Weighing and Feeding Guide,* the WT10 catalogue, and collection of case studies present a full complement of information.

- contribute to additive ratio systems in the concentrator area
- record inventory of concentrated product, and of loadout control and loadout records

Solids flowmeters and belt scales are used in mine backfill systems where concrete fills exhausted cavities and reinforces support columns within the underground mine. Solids flowmeters and belt scales are also used within smelters as production monitors.

The aggregates industry uses belt scales for recording inventory going to stock piles and for loadout control, for recording material transfer to trucks and rail cars, and for managing blending control for special aggregates mixes.

Cement

In the cement industry, weighing systems are used in numerous locations:

- Belt scales in the limestone quarry for feed rate control to primary and secondary crushers and for control and recording of loadout to ships and trucks.
- In the raw material preparation processes, belt scales record raw materials inventory as it enters the storage area. Weigh-feeders are often used for feed rate control of raw materials into the raw mill grinding area.
- In clinker production, solids flowmeters monitor the preparation and control of the raw meal powder cement fed to the kiln. They also monitor pulverized coal when it is used as a fuel. In some instances, belt scales monitor and control rubber tires added to the kiln as a secondary fuel.
- In the finish mill, weighfeeders control and measure the clinker cement to additives ratios as the various components are added. Solids flowmeters monitor the dust separator outputs to ensure finish mill efficiency. Solids flowmeters and belt scales control and record finished cement blending and loadout to rail cars, ships, and tanker trucks.

Food and beverage

Belt Scales:

- measuring raw vegetables in loading systems and for inventory control

- doing inventory control, recording grain input into a grain terminal and grain processing factory, as well as monitoring other plant raw materials such as granular sugar and sugar beets

Weighfeeders (or weighbelts):

- weighing raw vegetables, processed vegetables, and potato chips and cut fries before and after the cookers

Solid flowmeters:

- monitor flow rates of grains and flour mill products
- monitor the processing of breakfast cereals

Belt scales, weighfeeders, and solids flowmeters are used for the blending of dry mash with additives or coatings in the production of animal feed.

Secondary markets

Everywhere product is moved, Siemens weighing can play a significant role in the control and measurement. Thus while the markets listed below may not have a dominant need for weighing equipment, they do have many opportunities for weighing.

Chemical (dry)

Belt scales play a limited role in the chemical industry, other than in the inventory and load out control of finished dry product such as plastic pellets, resin powders, sodium nitrate, sodium chloride, and other dry additives.

Metals

Weighfeeders are used in the blending of raw materials that feed the furnaces in iron and steel production.

Environmental

Belt scales operate in recycling plants (plastic, aluminum cans, paper) to keep track of plant inventory and monitor processing.

Water/wastewater

- Belt scales monitor and control sludge fed to furnaces in sewage treatment plants.
- Small weighfeeders and solids flowmeters control the addition of treatment chemicals in municipal water supply systems.

Energy

- Belt scales monitor and record the transfer of coal from mines to coal-fired power plants, ship-to-coal yard storage, and the blending of coal types for optimum burning and control of emissions.
- Weighfeeders are used to add lime for emission control.
- Belt scales and solids flowmeters are used to blend burner sludge; fly ash, and lime for strip mine backfill; as well as in the loadout of fly ash for shipment.

Pharmaceutical

Weighbelts are used for the production monitoring of mass produced tablets and capsules.

Oil and gas

Solids flowmeters are used for truck loadout of rock trap filler, material used with cement to create concrete seals around oil well shafts.

Pulp and paper

Belt scales and solids flowmeters are used in inventory and processing control of wood chips and sawdust.

Mining, aggregate, cement

Application one: efficient mine backfill

Tailings are what remains of the ore after the mineral extraction process.

Once sections of a mine have been depleted, they are backfilled with a variety of material. Mine backfills prevent cave-ins by filling voids and air blasts, improving structural support, re-contouring waste piles, and reclaiming high walls by increasing stability. Cement is a popular component of backfill material and in some areas it accounts for as much as 10% of the region's total cement consumption. For mining companies, backfill cement can make up as much as 5% of their operating costs.

To offset the cement cost and still maintain the quality, strength, and consistency of the backfill, other materials are often added:

- flyash
- pozzolana[3]
- slag
- gypsum
- lime

---

[3] A fine, sandy volcanic ash.

These ingredients are mixed with water, sand, and other aggregates, but most commonly with the mine's tailings. The challenge is to increase the strength and quality of the backfill to ensure safety while simultaneously minimizing material and operating costs.

Conditions

Recipes for backfill differ depending on its purpose for the individual batches. The ratios at which ingredients are blended need to be closely monitored to ensure the correct recipe is followed.

- lower grade backfill can be used in non-critical applications
- higher grade backfill with more tensile strength is for areas of high load

Solution

To ensure the proper mix, Siemens installs the SITRANS weighfeeder with a Siemens Milltronics BW500 integrator to regulate the addition of the binding material. Stored in silos, the materials are generally dispensed onto the weighfeeder from the silo bottom with a rotary air lock feeder.[4]

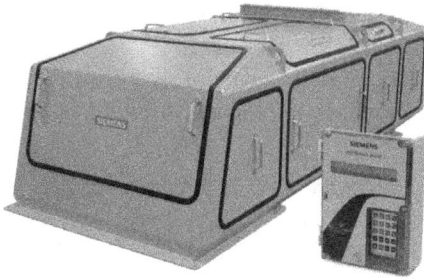

The weighfeeder is usually enclosed, with a dust collector port to control the fine material dust. It weighs the material as it is conveyed to the mixing tanks where it is combined with the other ingredients.

The rate of material flow is compared to a setpoint in a control loop, and the speed of the rotary valve and the speed of the weighfeeder's belt are varied in unison to deliver a controlled accurate flowrate of material to the mixing process. The BW500 Integrator has the ability to provide a fully functioning dual loop PID controller as well as be programmed for batch control at the same time.

---

[4] The rotary airlock is preferred because of the dusty nature of the fine materials.

Benefits

The Siemens SITRANS weighfeeders and BW500 Integrator accurately weigh and consistently control the flow of the expensive binder materials into the mixing process to maintain the backfill quality and plant efficiency while minimizing the operating costs. Weighfeeders are a rugged, low maintenance, cost-effective solution that will provide years of trouble free, accurate weighing and material feeding.

Application two: production rate measurement in a potash mine

Potash (potassium carbonate) occurs abundantly in nature and is the seventh most common element in the earth's crust. It has numerous uses:

- in the manufacture of glass and soap
- as soil fertilizer because it improves water retention, yield, nutrient value, taste, colour, texture, and disease resistance in food crops
- as nutrient for fruit and vegetables, rice, wheat and other grains, sugar, corn, soybeans, palm oil, and cotton

Although potash is used in over 150 countries, it is only produced in 12. Demand has increased for food, fiber, and animal feed, supplemented by an increased demand for biofuels. The current potash market of 50 million tons is projected to grow at a rate of 3-4% per year.

This global producer and distributor of potash and phosphate for industrial use operates 16 phosphate rock mines and plants, five potash production facilities, and one nitrogen production facility.

Conditions

- weigh raw potash ore at Canadian mining location
- high accuracy required to determine mine production rates
- 48-inch belt conveyor suspended from the mine ceiling carrying 2,800 short tons per hour at 600 feet per minute

The suspended conveyor made the belt scale design a challenge, and accuracy was influenced by the dynamic forces created by the conveyer support method.

Solution

Siemens installed the following system:

- Milltronics MSI belt scale
- Milltronics BW500 integrator
- SITRANS WS300 speed sensor attached to a six-inch bend pulley

The Milltronics MSI provides accuracy of ±0.5% of totalized rate. The stainless steel parallelogram load cells with their 300% overload protection provide instant response to vertical loading and eliminate any dynamic forces generated by the horizontal movement of the belt. The strong static beam support also compensates for any dynamic forces created by the suspension of the conveyor. With belt speeds up to five meters per second (1000 feet per minute), the MSI and the BW500 handle mine production rates quickly and efficiently.

Benefits
The Siemens belt scale system provided immediate reliable weighing results, increasing accurate production measurement. Other benefits were also immediately realized:

- unlike the multi-idler systems the company was using, the single-idler MSI and components cost up to 25% less
- reduced shipping weight resulted in lower shipping costs
- the compact MSI did not require special equipment to move or install
- down time is significantly lower during installation because the belt scale is designed to Conveyor Equipment Manufacturer Association (CEMA) standards, allowing for simple drop-in installation with no conveyor modification required

The Milltronics MSI/BW500 package provided this company with an economical and reliable solution for measuring mine production. A long-time customer of Siemens Milltronics, they continue to place orders for Siemens weighing and level products.

Application three: custody transfer with NTEP belt scale

Custody transfer applications require a certified accuracy so that both parties can rely on a guaranteed understanding of the transfer value of the goods. The Siemens MMI/BW500 system has National Type Evaluation Program (NTEP) certification of conformance for weighing and measuring devices from the National Conference on Weights and Measures (NCWM).

The MMI/BW500 system also has Electronic Indicator/Totalizer approval from Measurement Canada, making it legal for trade in Canada. As well, there are similar approvals for the European Union (OIML[5]).

---

[5] International Organization for Legal Metrology. Go to www.oiml.org for additional information.

Conditions

A major aggregate producer in the USA needed the capability to load rail cars with material from a conveyor belt while simultaneously creating a custody transfer transaction or point of sale. The producer also needed to maximize load-out efficiency and be efficient with the time required to fill each car. The state Weights and Measures Board would certify the belt scale to the National Institute of Standards (NIST) Handbook 44.

The following had to be considered:

- an invoice or ticket to legally document the transaction was required from the system
- certifiable belt scale was mandatory
- 122 centimeter (48 inch) wide belt conveyor
- conveyor operated from 3,000 to 3,500 tons per hour
- rail car loaded in about three minutes

Solution

Siemens installed the MMI/BW500 NTEP system[6] providing ±0.25% accuracy over the totalized range – it has passed all the laboratory and field testing required for NTEP approval. The system was installed and calibrated to certifiable standards in accordance with the guidelines outlined in Handbook 44.

Milltronics MMI is a two-idler belt scale ideal for custody transfer and for fast-moving belts, short idler spacing, and light or uneven belt loading. Its stainless steel parallelogram load cells have 300% overload protection and provide instant response to vertical loading, eliminating any influences generated by the horizontal movement of the belt such as those caused by high winds. With belt speeds up to five meters per second (1000 feet per minute), the MMI handles loading of the rail cars fast and effectively.

Benefits

The Siemens MMI/BW500 system costs 15-20% less than the standard multi-idler configuration and comes in at only a third of the weight so shipping costs are lower as well.

It's easily installed without heavy equipment and with less down time needed for installation. Its stainless steel triple beam load cell design is ideal for harsh conditions. Furthermore, fewer idlers and a reduced number of surfaces for product buildup lowers maintenance costs.

---

[6] Milltronics MMI belt scale, Milltronics BW500 integrator, SITRANS WS300 speed sensor, and static test weights along with a power conditioner, material flow alarm, and ticket printer.

The Siemens MMI/BW500 NTEP package brought this aggregate producer customer an economical and reliable solution for certified custody transfer, keeping its sales numbers and costs on track.

Application four: efficient loadout for aggregates

A major UK aggregate company with more than 70 operations, including crushed rock, sand, and gravel quarries; as well as a growing network of secondary and recycled aggregate depots, needed to improve its loading accuracy at a quarry in Leicestershire. Inaccurate loading has three major consequences:

- accurate weights ensure maximum permissible weight loads for transport
- under loading is inefficient and increases transportation costs as more runs need to be made
- over loading is expensive and time consuming because excess material must be unloaded and is then often unrecoverable as finished product

Conditions
- 5,000 tons of finished product every daily
- 3,000 tons are loaded into railway cars, each holding approximately 67 tons
- 2,000 tons are transported by truck, each delivery truck holding approximately 26 tons

Solution
Siemens installed two MSI (Milltronics Single-Idler) belt scales, one for each conveyor. The scales were simply dropped in, each fastened with four bolts, existing idlers were attached to the MSI's dynamic beam and accurate idler alignment procedures were followed.

Siemens installed Milltronics BW500 Integrators and speed sensors to calculate flow rate, total weight, belt load, and belt speed.

Benefits
The accuracy of the MSI system was well within the company's 0.5% stated specification so they modified the loading system to fill the railcars directly from the information provided by the MSI scale. When higher accuracy was required later on, a second MSI scale was added to create a multi-idler system (MMI).

The success of this installation means that all road and rail loadouts are now permanently controlled by MMI systems (20+) on each conveyor for inventory management and control. The MMIs are calibrated every three months, and accuracy readings of 0.1% against materials tests are common.

Accurate weighing helps ensure maximum loading of railcars and trucks.

Application five: accuracy in cement

A Chinese cement company with a cement production line of 5000 tons a day has a loading site with two wharves on the Yangtze river delta. Each wharf has a conveying belt and a belt scale for custody transfer and trade settlement. One belt scale is used for coal input and the other for cement output. Accuracy on these is problematic because the long belts are affected by environmental conditions.

Conditions
- long conveyor belts: coal input belt is 1100 meters and cement output is 600 meters
- heavy wind and other aggressive weather conditions
- coal input belt addressed first. It has three belt splices, runs at two meters per second, and has a flowrate of 400 tons per hour.

The site faced long-standing performance challenges with its existing belt scale systems, including low accuracy, poor stability, and high maintenance costs. They wanted to increase the flowrate to 700 tons per hour with 0.5% accuracy.

Solution
Siemens installed the Milltronics MMI double-idler belt scale with a BW500 Integrator and the WS300 Speed Sensor. The MMI comprised two MSI belt scales in tandem, creating a superior multi-idler belt scale system offering ±0.25% accuracy. With trade approval certificates,[7] the MSI/MMI offers outstanding accuracy and repeatability for weights and measures applications globally.

The overall system accuracy was tested three times with 65 kilo per meter test chain calibration.

Benefits
Since the installation of the Siemens belt scale package, the zero point offset is less than 0.05%. The tests resulted in a repeatability factor of better than 0.15%.

---

[7] OIML, MID, NTEP, and Measurement Canada.

The MMI belt scale's combination of accuracy, ease of installation, commissioning, and maintenance are unparalleled in today's industrial world. For this customer, the MMI belt conveyor package provides a cost-effective and low-maintenance solution.

Application six: OIML belt scale keeps aggregates on track

At the UK branch of one of the world's largest suppliers of heavy building materials to the construction industry, a yearly average of 750,000 tonnes of aggregates, asphalt, and cement and 80,000 cubic meters of ready-mixed concrete is delivered by rail, water, and road. Accurate measurement of this volume is crucial as any variation leads to significant revenue loss. The company required accurate loading data for material filling hopper rail cars.

Conditions
- hopper rail cars filled at a rate of 700 tonnes per hour
- loaded car holds 66 tons
- 900 millimeter conveyor belt travelling at 1.6 meters per second
- fill time should be under six minutes

Under loading is inefficient and increases transportation costs. Overloading is expensive and time consuming as excess material must be unloaded and is often unrecoverable.

Independent verification of the weigh accuracy needed to be documented for both the rail operator and for the receiver of the dispatched material. The company requested that the installation meet National Measurement Office (NMO) custody transfer certification.

Solution
Siemens provided an OIML approved Milltronics MMI belt scale system which includes the following:

- Milltronics MMI belt scale
- Milltronics BW500 integrator
- SITRANS WS300 speed sensor
- static test weights for calibration

Further equipment required for OIML approval to verify and record the transaction:

- out-of-range indicator
- non-resettable remote totalizer
- ticket printers are required under the OIML approval conditions

139

Benefits

The two-idler Milltronics MMI offers high accuracy (±0.25%) over the totalized range. The stainless steel parallelogram load cells with 300% overload protection that provide instant response to vertical loading, eliminating any influences generated by the horizontal movement of the belt. With belt speeds up to five meters per second, the MMI handles loading of the hopper wagons fast, efficiently, and accurately.

Agriculture

Application one: light loading

Light loading presents its own challenges. The material may not exert even pressure on the belt and it is subject to spillage from the movement of the belt.

A solids bulk handling manufacturer needed to measure output on a tobacco conveyor. Output needed to be totalled to ensure product was not lost or removed during the process.

Conditions
- small conveyor with the pulley centers at 2430 millimeters (95 inches)
- limited installation space
- low bulk density of material but it can be in large units
- material, ranging between 6-28 pounds per cubic foot, is typically fed onto the conveyor from a hopper
- flat belt profile
- 1000 millimeter wide belt
- material containment skirt boards 400 millimeters (15.75 inches) high

Solution

Siemens installed the Milltronics MLC belt scale, a special weighbridge version of the MSI specifically designed for light loads and flat profiled belts. The scale comes complete with an idler to ensure that dead loads are kept to a minimum. The compact MLC was mounted directly to the conveyor structure, ensuring a rigid connection. The load cells are mounted on a channel beam spanning the entire width of the conveyor, keeping them parallel and providing optimum weighing performance.

The client works with SIMATIC PCS 7, the Siemens control system, so they installed a SIWAREX FTC module for the integrator function and a TASS speed sensor to give belt speed data.

Benefits
The Milltronics MLC belt scale provides continuous total indication for the conveyer end user. Furthermore, high accuracy and consistent reliability ensure a low cost of ownership and the most efficient use of material.

The MLC and the TASS speed sensor are also stainless steel as they are washed regularly to avoid contamination.

The MLC belt scale is rated at ±0.5% of totalization over a rate range of 25-100%. The conveyor manufacturer runs the tobacco at a maximum of two tons per hour and 0.1 meters per second.

Application two: blending and batching for feed production

A company in The Netherlands produces milk replacement livestock feed which it distributes worldwide. The company needed to upgrade the blending and batching system to an integrated solution at one its main facilities. It turned to Siemens who worked with the company's engineering team and a system integrator to design a solution.

Conditions
The facility produces a powder with a high fat content from raw dairy materials. The semi-finished products are made using a computer-controlled process involving mixing raw materials in liquid form, pasteurization, homogenization and spray-drying. The semi-finished goods are then transported to another manufacturing unit where they are stored in silos until they are processed into batches of finished product.

Traditionally batches were as large as 300 tons, but the need for increased production and smaller batches also defined the requirement for a new blending and batching system with a higher accuracy. This situation had to be integrated into the ERP (Enterprise Resource Planning) environment of the plant.

- twenty silos transporting ingredients to the in-line blending process blending system comprising twelve stand-alone weighfeeders
- each weighfeeder was pre-fed by a bunker with a rotary feeder providing volumetric discharge
- there was no integration with the other operations in the plant
- downtime had to be as short as possible to reduce productivity losses

Solution
Siemens offered an integrated solution with fourteen high accuracy
SITRANS WW200 Series SD (sanitary duty) weighfeeders integrated
into Simatic PCS 7. Simatic IT joins the new batching and blending
system with the ERP environment. Siemens supplied the WW200
Series weighfeeders complete with drives, load cells, and motors.

The frame of the WW200 series weighfeeder is sturdy and rigid,
ensuring stable, repeatable results while maximizing resolution and
weighing accuracy. The WW200 series is designed for the food
industry where high-pressure wash down is required and meets all
USDA and FDA requirements.

- the weighfeeders blend and batch fourteen different products simultaneously
- weighfeeders are fed with ingredients from the bunkers,
  weighing the batches with an accuracy of 0.5%. The speed of
  the belt and the belt loading determine the correct dosage. To
  avoid a high turndown in belt loading, the material depth on
  the belt is adjusted automatically for the different materials.
  - the shear gate level positioner is powered by a Siemens
    Posmo multi-step drive
  - SIWAREX FTC modules accurately determine the feed rate
  - SIMATIC PCS 7 controls the belt speed automatically accord-
    ing to the process readings and the established set points
  - SIWAREX U, the versatile weighing module, provides sim-
    ple weighing and force measuring tasks

The bunkers are supported by SIWAREX load cells that measure filling levels. The weighfeeders have an extra weighdeck at the weighfeeder inlet area to measure product left on the belt when the

weighfeeder is stopped. Both the bunker and the extra weighdeck load cells are connected to SIWAREX U processor modules so product is continually measured, making for faster programming time between batches and reducing product losses to zero.

The bunkers and the weighfeeders are connected through the SIWAREX FTC and SIWAREX U processor modules by linking SIMATIC PCS 7 via PROFIBUS DP which is linked through PROFIBUS DP to a SIMATIC S7-400 system.

Benefits
The compact design of the weighfeeders made it easy to fit them into their positions. The shutdown of just seven days and loss of only one small batch in the first test run was a significant benefit to the company.

The weighfeeder load cells are externally mounted for easy access and maintenance.

The recipes for the batches are made with a special software, and are loaded into SIMATIC IT, which subsequently forwards them to SIMATIC PCS 7 to control the batching system. This replaced the original manual recipe programming which had to be done for every batch and saves an enormous amount of time.

The operator can now freely configure the automation solution – including the scale application – and run diagnostics on the load cells and weighfeeders program, implementing a weighing and proportioning system that the company can adapt to changing operational requirements.

Starting and stopping the weighfeeders, as well as the changes from one product to the other, are now simple as the entire control system was changed to SIMATIC PCS 7, complete with all its functions.

The result is increased efficiency, flexibility, and production capacity.

Summary

These applications are only a few of the thousands that Siemens has set up in the last thirty-five years. Weighing opportunities lie everywhere and Siemens is committed to providing the most accurate and durable weighing systems suited for any application.

Clients throughout the world rely on Siemens equipment to service their customers, and Siemens looks into the future to provide them with the latest innovations in weighing technology and communications. When someone asks for the best in performance – the answer is always Siemens.

Appendix A

Glossary

Knowledge is of two kinds. We know a subject ourselves, or we know where we can find information upon it.[1]

Analog to digital (A/D) converter a discrete electronic circuit, or an integrated circuit (IC), that converts an analog signal to a digital signal.

Approach idlers A1, A2, etc conveyor idlers the belt rolls over before riding on the scale idler.
- A1 is the last one before the scale idler
- A2 is the second last
- A3 is the third last

Balance a weighing device comprising a rigid beam horizontally suspended by a low-friction support at its center. Identical weighing pans are hung at either end; one pan holds an unknown weight to be determined while the other pan has known weights added to it until the beam is level. The unknown weight is the same as the sum of the weights added to the other pan.

Batching a weighing measurement function that counts totalizer inputs (weight readings) and triggers an event once the required value is reached.

Belt scale a weighing instrument specifically designed to weigh material transported on a conveyor belt in motion (aka belt weigher).

Belt plough a diagonally mounted device which forces the conveyed material off the side of the belt conveyor during operation. Often used to empty the belt or to change the belt loading profile so that the loading is efficient.

Belt speed the lineal speed of travel of the belt, usually in terms of meters per second or feet per minute.

Belt take-ups keep consistent conveyor belt tension, managing slack belting created by belt stretch and temperature elongation. Most common types are:
- Vertical gravity
- Horizontal gravity
- Horizontal screw types

Bend pulley a pulley in a belt conveyor that forces a bend in the "clean side" of the return belt to take up belt slack or to establish a drive pulley for a belt speed sensor.

Bin a general term for any chute with additional width and depth for temporary storage.

Bogey wheels small metal wheels that run on and are supported by metal tracks, much like the wheels on a railroad car.

---

[1] Samuel Johnston. *The Life of Samuel Johnston,* John Boswell. 1791.
The authors also appreciate www.Wikipedia.org for its vast reservoir of information and assistance clarifying some of the terminology.

Carrying idlers conveyor idlers that support the belt and its loading.

Centripetal force the force that makes a body follow a curved path: it is always directed orthogonal to the velocity of the body and toward the instantaneous center of curvature of the path.

Chute is a vertical or inclined plane, channel, or passage through which objects are moved by means of gravity.

Control gate controls the depth of load and the profile of material on a belt conveyor or a weighfeeder.

Conveyor idlers single, dual, triple, or quadruple roller devices that are rigidly supported by the belt conveyor structural frame. Because of the bearings supporting the rollers, the idler allows the conveyor belt to travel with the rollers turning "idle."

Coriolis effect a deflection of moving objects when they are viewed in a rotating reference frame. The object does not shift from its path; it just appears that way because the reference system has shifted.

CPU the Central Processing Unit of a computer system.

DCS a Distributed Control System has one central control computer working with other local controllers to control plant operation.

Diamond port knife gate the gate valve opening creates a rectangle-shaped opening (rotated 90 degrees), providing a better linear cross-sectional area for the gate opening than round port or v-port opening valves.

Digital to analog (D/A) converter changes a digital signal to an analog signal using a discrete electronic or integrated circuit (IC).

Digital format analog signals can be held and transmitted by a physical variable such as current, voltage, or pressure and are digitized as a series of 1s and 0s. This conversion depends on the type of digital format being used as there are several available. The difference between them lies in the degree of accuracy of the represented value. All the digital formats can be easily manipulated and processed, permitting instrumentation companies to build in functions to clean up raw data of noise and allow for easy filtering and totalizing of values.

Dosing controlling the blending or mixing of materials to a predetermined ratio. Also considered as flow rate control.

Dry weight material weight when all significant moisture is removed. Usually based on calculations using non-dry material weight and moisture meter measurements.

Flexure bearing a typical flexure bearing is just one piece joining two other parts and made of a material which can be repeatedly flexed without fatigue. It provides linear movement in direct proportion to forces applied. They have range of motion limitations and are not recommended for bearings that support high loads.

Gravimetric feeding a feeding system that derives material flow rate by using a transducer (load cell or a LVDT) to measure the gravitational force exerted by material on a conveyor belt and a signal proportional to the speed of the conveyor belt. Generally, they perform better than volumetric systems.

Gravimetric weighing weighing of material by measuring gravity forces by these systems: belt scales, belt weighfeeder, weighbelts, rotary weighfeeders, and loss-in-weight systems.

Head pulley the main pulley at the discharge end of a belt conveyor.

Hopper a general term for a chute with additional width and depth for temporary storage.

Horizontal gravity take-up a mechanism pulling the tail pulley horizontally using cables applied to weights over pulleys. This arrangement takes up any belt slack and provides more consistent belt tension.

Horizontal screw take-up a mechanism pulling the belt horizontally and keeping it taut despite tendencies for the belt to slacken – an inferior method as there is no variable compensation applied to the belt tension.

Hysteresis when a system cannot produce repeatable input related results or repeat results for a given condition.

Impact idler a conveyor carrying idler that is built to withstand the impact of material dropping onto the belt.

Inferred weighing weighing material measuring influences other than gravity, including impact, centripetal, centrifugal, and Coriolis forces, includes solids flowmeters.

Linearizer refers to a programmable integrator feature that allows for compensation of non-linear (load) readings.

Load cell a transducer that converts mechanical force into a usable signal (electrical, hydraulic, or pneumatic). The strain gauge load cell is the most common and typically uses four strain gauges connected in a Wheatstone Bridge configuration to produce an electrical signal proportional to forces applied. A second type, the LVDT, measures movement and produces an electrical signal proportional to the same.

Loss-in-weight weighing of material in bulk by deriving a flow rate calculation as the hopper reduces in weight when a mechanical feeder discharges material from the hopper.

LVDT Linear Variable Differential Transformer, an electrical transformer that measures linear displacement.

Mass often used as a synonym for weight when at rest.

Metrology the study of weights and measures, including units of measurement.

PLC Programmable Logic Controller used in the control of process and sequential operation of equipment.

Pre-amplifier (P/A) a discrete electronic circuit or integrated circuit (IC) used to increase a signal level, such as from millivolts to volts. The output of the circuit needs to be directly proportional to the input.

Quadrature output a term for the two outputs on an incremental rotary encoder (a.k.a. quadrature encoder or a relative rotary encoder). Quadrature refers to how the two output wave forms are 90 degrees out of phase. Interfacing logic checks the signal created by shaft vibration, verifying that there are at least two out-of-phase pulses, confirming shaft rotation.

147

Radar level Level detection using transmission and time measurement of reflected microwave radar pulsations.

RAM Random Access Memory is used to temporarily store digital information by the microprocessor.

Receiver an instrument that receives a signal.

Retreat idlers R1, R2, etc. the idlers the conveyor belt goes over after riding on the scale idler.
- R1 is the first after the scale idler
- R2 is the second one
- R3 is third one

Return idlers belt conveyor idlers, usually flat rollers, which support the return portion of the belt on the underside of the belt conveyor.

ROM Read Only Memory, used to permanently store digital information, including the microprocessor operating instructions.

Sealing strip a strip of material, usually rubber, that keeps material from escaping from between the skirt boards and the conveyor belt.

Scale idler S1, S2, etc. the weighing idlers supported by the weighbridge.

Scale or weigher an instrument, device, or a machine for weighing.

Shearing the extraction of de-aerated material from the bottom of a bin or silo, often done with a special belt conveyor, screw conveyor, table feeder, rotary feeder, or apron feeder.

Silo structure for storing bulk materials.

Skirt boards vertical walls running parallel to the conveyor belt in the belt in-feed area to keep material from falling off conveyor.

Snub pulley a pulley on a belt conveyor used to create enough wrap around on the conveyor drive pulley to help prevent slippage, generally located close to the head pulley.

Strain gauge thin metallic foil pattern with a specific electrical resistance that changes proportionally to any force or strain applied to the gauge.

Span the range of operation from zero to 100% load.

Stacker conveyor a belt conveyor (usually portable) with a variable incline adjustment so that its through-put creates a stack without overly disturbing the distribution of fine and coarse components in an aggregate mix.

Surge bin a chute with additional width and depth for temporary storage.

Tail pulley the main pulley at the feed end of a belt conveyor. It usually rotates in reaction to belt movement; however, it can sometimes be used as the belt travel driver.

Totalization the running total of material conveyed.

Transmitter an instrument that transmits a signal.

Tripper conveyor a belt conveyor offering various feed points through a mechanism that moves along the discharge area, often used to discharge material into a series of bins or silos.

Truck scale a weighing system that weighs a vehicle statically. The truck is driven onto a weighbridge where it is weighed when it stops.

Trunnion a pin or pivot on which something can be rotated or tilted.

Vertical gravity take-up a weighted, free-traveling pulley that vertically moves with gravity to provide more consistent belt tension by taking up belt slack.

V-notch (V-port) a V-shaped opening on an gate valve, created as the gate opens.

Volumetric feeding a feeding system using a device to deliver a specific, predictable and repeatable volume of material with each revolution or cycle. These devices can be varied in speed to deliver a specific and reliable rate of material feed as long as the material is uniform in size, mass, moisture content, and bulk density and is not prone to bridging between surfaces or sticking to surfaces.

Weigh determining the force exerted by gravity using balance, scale, or other mechanical device.

Weighbelt a belt conveyor designed to transfer material at a constant belt speed, or material depth, with special emphasis on weighing the material during transfer.

Weighfeeder a belt conveyor designed to feed at various rates by varying the belt speed or material depth with special emphasis on weighing the material during transfer.

Weighbin a bin weighing its contents using load cells.

Weighing idler the weighing idlers supported by the weighbridge (same as a scale idler).

Weight a measurement of the gravitational force on a body using a known reference.

Weights and measures the laws, acts, rules, governing bodies, agencies, or standards dealing with the science of metrology and the measurement of physical quantities.

Index

THIS TITLE IS FROM OUR MANUFACTURING COLLECTION. OTHER TITLES OF INTEREST MIGHT BE...

Announcing Digital Content Crafted by Librarians